RAND | NATIONAL DEFENSE RESEARCH INSTITUTE

Assessment of Surface Ship Maintenance Requirements

Robert W. Button, Bradley Martin, Jerry M. Sollinger, Abraham Tidwell

Prepared for the United States Navy

For more information on this publication, visit www.rand.org/t/RR1155

Library of Congress Cataloging-in-Publication Data is available for this publication.
ISBN: 978-0-8330-9253-3

Published by the RAND Corporation, Santa Monica, Calif.
© Copyright 2015 RAND Corporation
RAND® is a registered trademark.

Support RAND
Make a tax-deductible charitable contribution at
www.rand.org/giving/contribute

www.rand.org

Preface

The Department of Defense is likely to face years of declining resources as the U.S. government grapples with fiscal challenges. At the same time, there is no evidence that the demand for U.S. forces, in particular naval forces, is going to decline. For its part, the planned replacement rate of ships in the U.S. Navy will result in an aging fleet; the cost of maintaining ships, especially in the surface fleet, will increase accordingly. Reduced budgets will only make surface ship maintenance more difficult. This will exert particular pressure on the parts of the Navy establishment supporting materiel procurement and readiness. This study looked in particular at surface ship maintenance to assess trends, determine the impact of a constrained fiscal environment, look in detail at some ship classes, and then look at potential mitigations for what appears to be a currently unsustainable maintenance path.

This research was sponsored by the Assessment Division (N81) of the Office of the Chief of Naval Operations and conducted within the Acquisition and Technology Policy Center of the RAND National Defense Research Institute, a federally funded research and development center sponsored by the Office of the Secretary of Defense, the Joint Staff, the Unified Combatant Commands, the Navy, the Marine Corps, the defense agencies, and the defense Intelligence Community.

For more information on the RAND Acquisition and Technology Policy Center, see www.rand.org/nsrd/ndri/centers/atp or contact the director (contact information is provided on the web page).

Contents

Figures

Tables

Summary

The Department of Defense is likely to face years of declining resources as the U.S. government grapples with fiscal challenges. These challenges affect every account, including those associated with surface ship maintenance and operations. At the same time, there has been widespread concern that surface ship materiel readiness is declining due to a high pace of operations and a sense that there have been many instances of deferred maintenance. The need to balance fiscal reality and a continued need for ready ships is likely to be an ongoing challenge.

What We Were Asked to Do

The Navy asked researchers from the National Defense Research Institute (NDRI) to accomplish three tasks. First, it asked us to determine the effect of near-term reductions in Operations and Maintenance (O&M) accounts on long-term fleet readiness, specifically Operational Availability (Ao) and Expected Service Life (ESL). Second, it requested NDRI to develop a maintenance requirement concept for each ship class that supports ESL but allows for some risk within the maintenance strategy. It also asked researchers to define the risks to Ao and ESL resulting from the new requirement. The concept should apply to multiple ship classes. Third, it asked researchers to recommend potential strategies to minimize negative effects to Ao and ESL and maintain the largest, most capable fleet possible. Note that the tasking did not request a complete explanation of a particular observed trend; both the study team and the sponsor concluded that this would require a larger effort with a larger amount of data than is currently available. Focus was on characterizing the trends, describing how funding changes might impact these trends, and then suggesting methods to possibly mitigate the impact.

How We Studied the Problem

We first examined the fiscal environment likely to face the entire Navy (including the part of the Navy that carries out surface ship sustainment). We examined the overall budget history, then the projected environment based on the caps placed on base-budget spending enacted in the Budget Control Act of 2011 and the short-term Bipartisan Budget Act of 2013. We also examined the trend of expenditure per ship on surface ships, plotting this across several ship classes through their lifecycles. The intent was to both show the degree of upward pressure on expenditure and the degree of constraint imposed by external factors. We then examined current practice for scheduling and prioritizing maintenance, first by describing the process, then by comparing the actual work performed against the notionally specified levels. We also attempted to account for the potential impact that would result from failures to complete required maintenance. We did this principally by looking at the operational and maintenance histories of forward deployed ships, with limited opportunity for availabilities and comparing them to ships stationed in U.S. homeports with the normal maintenance opportunities.

Finally, we developed and applied heuristics to enable prioritization of work within Selected Restricted Availabilities (SRAs) and also between SRAs and Continuous Maintenance Availabilities (CMAVs).[1] This required assessment of Ships Work Line Items across factors such as urgency, predictability, and long-term cost.

What We Found

Maintenance has historically been a fixed fraction of total O&M budgets. Expenditures per ship on surface vessels have varied over the years, not showing a particular trend of steady upward movement. Indeed, in the early 1990s, these expenditures actually declined as the Navy decommissioned whole classes of old ships while building new ones. Since 1998, however, the expenditures per ship have gone up steadily, with an exponential function being the best fit for describing cost per ship through this period. This runs directly into the known legislative environment, which will in fact impact available resources. Annual shortfalls now amount to $8 billion between requested budget and future possible budgets if legislation does not change. Indeed, pressure on maintenance budgets is expected to grow as the effects of sequestration are felt, with no expected diminishment in the demand for naval forces. Although the Navy has historically managed maintenance costs by retiring old, cost-intensive platforms, pressure to maintain the force size, coupled with no new ship classes projected in the current shipbuilding plan, will likely preclude such relief in the next ten to 15 years and certainly not in the five-year Future Years Defense Plan time frame.

Even absent sequestration, continued growth in per-ship maintenance cost is likely unsustainable, at least at the rate seen in the last 15 years. At the rate seen, maintenance would either become a larger component of the O&M budget or come at the expense of new construction or modernization or require deferral. Maintenance choices, now being made, will become ever-more critical. At a minimum, within availabilities, choices would need to be made between activities that favor immediate to near-term Ao at the possible expense of longer-term maintenance to protect ESL.

Although we initially attempted to account for reasons behind particular variations among ships, we found that this was a far more complicated problem than existing data would allow us to answer. We did not attempt to account for the differences between ships or explain all the factors that might be impacting a particular trend line. Potential reasons range from training shortfalls to issues with the maintenance infrastructure to deficient manning models. The Navy's Surface Maintenance Engineering Planning Program (SURFMEPP) is a relatively new initiative that is collecting ship-level detailed data, which should prove valuable. In many

---

[1] An *availability* is a dedicated maintenance period. There are three basic types of maintenance availabilities: Chief of Naval Operations (CNO) Availabilities, Non-CNO (TYCOM) Availabilities, and and New Construction (SCN) Availabilities. CNO Availabilities are relatively long and are scheduled with CNO approval. CNO Availabilities include SRAs, which are discussed later. Non-CNO Availabilities include Continuous Maintenance Availability (CMAV) periods, which are also discussed later. SCN availabilities include Post Shakedown Availabilities. SCN availabilities are not discussed here.

cases, however, understanding why a particular availability escalated in cost would require on-site interviews with the ship's personnel, maintenance supervisors, and the overall chain of command. This would likely be very worthwhile work, but it rose beyond the sponsor's expectations for the study and is not necessary to answer some of the major issues the study did uncover.

A look at the DDG-51 class suggests a mismatch between what the Navy claims is essential for maintenance and what it has historically spent, both in individual years and cumulatively. The Navy publishes Technical Foundation Papers (TFPs) to state the level and kind of maintenance that needs to be performed at regular intervals within successive availabilities, leading to an expected cumulative level of maintenance. We examined in detail the cumulative maintenance levels for DDG-51s and compared them to the levels specified in the TFP. As the Navy-provided chart below (Figure S.1) indicates that the Navy is not in general funding to the TFP level.

It is not yet possible to assess the effect this mismatch has had on the class—the history is still being written—but there is doubt about either: (1) the validity of TFP requirements or (2) the Navy's commitment to actually carrying out the maintenance stated in the TFP. It appears that the Navy will need to consider alternatives to the TFP process as it formulates requirements and resource plans.

Maintenance requirements, as stated in TFPs, do not consider risk in its various forms. They also do not consider the complex issues, including cost, associated with deferring maintenance. They do not consider the net value of maintenance (the duration and degree of improved reliability as a result of maintenance). All of these considerations could contribute in planning more cost-effective ship maintenance.

Figure S.1
DDG-51 Flight I/II Cumulative Man-Days per Hull by Age

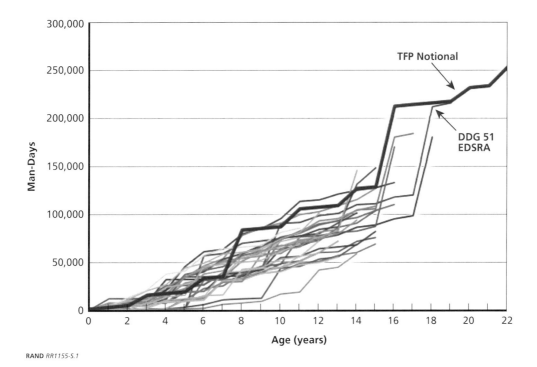

We also looked at other selected ship classes to evaluate the effect of deferrals, specifically to determine if there is a cumulative effect from deferring maintenance that might not yet be observable with guided missile destroyers (DDGs). The individual histories of ships affected the results. Individual cases suggest that maintenance deferrals, however, exact an extremely high premium that drives ship cost up in ways inconsistent with the need to contain costs. Ships of similar age and operating histories whose major difference is basing histories—with the attendant effects on maintenance—can show dramatic differences between the overall costs to maintain. Any maintenance construct needs to understand and budget for the high cost of deferral or devise mitigations for cases where deferral is inevitable.

Actions the Navy Can Take

The Navy has attempted to deal with these issues through organizational changes—such as SURFMEPP—and these will likely have a positive influence eventually. But the fundamental decision will still rest at the fleet- and type-commander level and will involve a weighing of operational, materiel, and long- and short-term factors.

We also looked at mitigation strategies, devising heuristic tools both to prioritize maintenance actions and to assign actions to differing types of availabilities. These are presented as a means to manage the consequences of limited maintenance funding or operational demands. We offer two maintenance approaches. One strategy prioritizes within SRAs and the other shifts maintenance from SRAs to CMAVs.[2] Although SRAs are relatively lengthy and well funded, the time and money available for maintenance in SRAs is limited. In the heuristic, maintenance priorities within SRAs are assigned based on safety concerns, the predictability of failures, the consequences of failures, and the future cost of deferring maintenance. If the maintenance required is relatively straightforward (i.e., does not require dry dock), and can be conducted with available personnel or can be broken down into manageable segments, then a CMAV can provide an efficient and cost-effective alternative to an SRA.

These heuristics are offered as a systematic approach to decisionmaking pending a more complete answer that we might expect from the several years of data collection that SURFMEPP will likely need to more completely answer the validity of maintenance requirements. They are not intended as relaxations of technical requirements, but as a means of managing competing requirements. They are, moreover, capable of being adapted to different decisionmaking venues and can accept alternate rule sets to the ones specified here. We recommend the Navy consider both alternatives as potential decisionmaking aids.

---

[2] SRAs are regular planned periods of maintenance, repairs, and upgrades. Ships in SRA are typically unavailable unless extreme conditions warrant. Work conducted during SRAs includes tank preservation, propulsion, and ship-system repairs and limited enhancements to various Hull, Mechanical, and Electrical systems. CMAVs are scheduled, relatively short (typically two to six weeks in duration) availabilities for surface ships normally scheduled once per non-deployed quarter during a period when the ship is in port.

Acknowledgments

Sean Critelli, Ken Munson, and Sean Ziegler provided valuable research support and insight into this study. We also thank Irv Blickstein, Paul DeLuca, and retired U.S. Navy Captain Michael J. Ford for their valuable insights in reviewing this report. The Chief of Naval Operations Assessment Division Readiness Branch was particularly valuable in its support. We offer especial thanks to U.S. Navy Commanders Harold DuBois and Neil Sexton, Stephen Williams, and Carlton Hill for their guidance and oversight.

Abbreviations

Ao	operational availability
BCA	Budget Control Act
BAWP	Baseline Availability Work Package
CASREP	casualty report
CG	guided missile cruiser
CMP	Class Maintenance Plan
CMAV	Continuous Maintenance Availability
CNO	Chief of Naval Operations
CONUS	continental United States
COMNAVSEASYSCOM	Naval Sea Systems Command
CSG	Carrier Strike Group
CSMP	Current Ship Maintenance Project
DDG	guided missile destroyer
DoD	Department of Defense
DSRA	Docking Selected Restricted Availability
EDSRA	Extended Docking Selected Restricted Availability
ESL	expected service life
FDNF	Forward Deployed Naval Forces
Flt	Flight
FY	fiscal year
HM&E	Hull, Mechanical, and Electrical
LCC	life-cycle critical
LPD	Amphibious Transport Dock
LMRS	Long-Range Maintenance Schedule

LSD	Landing Ship Dock
MMBP	Maintenance and Modernization Business Plan
MMPR	Maintenance and Modernization Performance Review
MRG	main reduction gear
MRS	Maintenance Resource System
NAVSEA	Naval Sea Systems Command
O&M	Operations and Maintenance
OFRP	Optimized Fleet Response Plan
OPNAV	Office of the Chief of Naval Operations
OPNAVINST	Office of the Chief of Naval Operations Instruction
OPNAVNOTE	Office of the Chief of Naval Operations Note
POM	Program Objective Memorandum
RCM	Reliability-Centered Maintenance
SCN	Shipbuilding and Conversion, Navy
SRA	Selected Restricted Availability
SUBMEPP	Submarine Maintenance Engineering, Planning and Procurement
SURFMEPP	Surface Maintenance Engineering Planning Program
SWLIN	Ships Work Line Item Number
T&V	Tank and Void
TBO	time between major overhauls
TFP	Technical Foundation Paper
TYCOM	type commander
VAMOSC	Visibility and Management of Operation and Support Cost

CHAPTER ONE

Introduction

Background

Surface ship maintenance has been a concern for a number of years, prompted by concern over episodes of poor performance on materiel inspections[1], growth in maintenance availabilities, and high-profile materiel failures. The failures of USS *San Antonio* (LPD-17) were so severe and pervasive that a flag-level investigation was commissioned to document their causes.[2] Although this report primarily focuses on the challenges of new ship introduction, a 2010 Fleet Review Board convened by then–Commander Fleet Forces Command Admiral John Harvey and chaired by retired Vice Admiral Phillip Balisle was a broader and very critical look at the entire force. Their findings concluded that surface ship materiel readiness was facing major challenges related to requirements identification, personnel training in understanding materiel conditions, and variable funding.[3] As reported in the unofficial media sources, the report held that several initiatives begun in the early 2000s to optimize manning or lower the burden of inspections and training requirements were at best ineffective and possibly counterproductive. Surface ship maintenance was, moreover, held to be underfunded for the materiel requirement. The short-term readiness and long-term sustainability of the surface force was held to be in some doubt. These findings have led to several efforts to both ensure more reliable funding and better identification of materiel requirements.[4] All are intended to promote Operational Availability (Ao) and achievement of Expected Service Life (ESL).

With all that understood, the Department of Defense (DoD) is likely to face years of declining resources as the U.S. government grapples with fiscal challenges. This demand will exert particular pressure on the parts of the Navy establishment supporting materiel procurement and readiness. The Budget Control Act (BCA) of 2011 established budget-enforcement mechanisms intended to reduce federal discretionary spending by more than $900 billion between 2012 and 2021. The BCA has broadly affected budgets within the U.S. Navy. In particular, surface ship maintenance budgets have been cut by an estimated 24 percent between fiscal year (FY) 2013 and FY 2015, with limited relief in sight. Worse, looking back over

---

[1] Christopher P. Cavas, "Two Ships Deemed 'Unfit' for Combat," *NavyTimes.com*, April 20, 2008.

[2] Department of the Navy, Commander, U.S. Fleet Forces Command, "Command Investigation of Diesel Engine and Related Maintenance and Quality Assurance Issues Abroad USS *San Antonio* (LPD 17)," 5830 / Ser N00/131, May 20, 2010.

[3] New Wars, "The Balisle Report and the Navy's Future Part 1," July 19, 2010.

[4] Michael Malone, "Surface Force Maintenance Engineering Planning (SURFMEPP)," project briefing given to the Surface Navy Association (SNA), January 17, 2013.

the last decade or more, a trend of rapid and unsustainable surface ship–maintenance cost increases is clear. Upward pressure for increased maintenance per ship seems destined to collide with the fiscal reality of legislatively constrained budgets.

Purpose

Prompted by these concerns, the Office of the Chief of Naval Operations (OPNAV), Assessment Division (N81) asked the RAND Corporation to accomplish the following tasks:

- Determine the effect on long-term fleet readiness, Ao, and ESL caused by near-term reductions in Operations and Maintenance (O&M) accounts.
- Recommend potential strategies to minimize negative effects to Ao and ESL and maintain the largest, most capable fleet possible.
- Develop a maintenance requirement concept for each ship class that supports ESL, but allows for some risk within the maintenance strategy. Define the risks to Ao and ESL resulting from the new requirement. *San Antonio*–class Amphibious Transport Dock ships and *Arleigh Burke*–class destroyers should serve as illustrative cases. The methodology should apply to multiple ship classes.

The study team executed these tasks, with one modification to substitute the *Whidbey Island* (LSD 41) and *Harper's Ferry* (LSD 49) class for the *San Antonio* (LPD-17), whose history was both too short and too troubled to be a reliable guide.

Note that the tasking did not request a complete explanation of a particular observed trend; both the study team and the sponsor concluded that this would require a larger effort with a larger amount of data than is currently available. Focus was on characterizing the trends, describing how funding changes might impact these trends, then suggesting methods to possibly mitigate the impact.

General Approach

The study carried out the following steps in our problem approach. First, we analyzed budget trends based on likely fiscal scenarios to show the likely path of maintenance funding. Second, we assessed trends in surface ship maintenance based on actual financial data derived from the Navy's Visibility and Management of Operation and Support Costs (VAMOSC) data base.[5] Thereafter, we assessed the degree to which the Navy is meeting its formal requirements (as described in Technical Foundation Papers [TFPs]) for the *Arleigh Burke*–class (DDG-51) destroyers. Since this class is still relatively new, we looked at older ship classes to attempt to assess the long-term costs of not performing maintenance, to include the Landing Ship Dock (LSD) class with guided missile cruisers (CGs) as a comparison. Then, based on the under-

---

[5]   VAMOSC is a management information system that collects and reports historic operating and support (O&S) costs. VAMOSC provides the direct O&S costs of weapon systems, some linked indirect costs, and related non-cost information, such as flying and steaming hours.

stood consequences of deferrals and maintenance limitations, we developed strategies to potentially mitigate the impact of these limitations.

Report Limitations

This report uses financial information to assess broad trends in surface ship maintenance across multiple ship classes. We assessed the difference between the stated requirement and maintenance actually budgeted and executed for a particular ship class for which there is both a well-developed maintenance requirement and a well-documented maintenance history. Execution, moreover, is not necessarily a statement of what was accomplished, but of what was spent in a given availability.

We considered a broad look at what might cause expenditures to increase or decrease, but this was ultimately not done by agreement between the research team and the sponsor before the project commenced. There are several possible reasons that costs might vary. Most possible reasons are interconnected to the point that establishing causation would be highly specific to circumstances, and data relating these factors is either not available or just beginning to be collected. Primary emphasis is on stated requirement, the degree to which funding has affected or will affect ability to meet the requirement, and possible mitigations when funding or schedule impacts ability to meet the requirement.

Report Organization

In Chapter Two, we describe trends in the Navy's maintenance costs and funding, both for the entire force and particular ship classes. In Chapter Three, we describe the expected fiscal environment for the Navy out to FY 2018. Chapter Four describes the Navy's current maintenance procedures for surface ships as well as the cost of deferring maintenance. Chapter Five describes two potential strategies to reduce the negative effect to Ao and ESL. Chapter Six provides a report summary. Appendix A describes additional concepts and techniques used successfully by Naval Sea Systems Command's (NAVSEA's) Submarine Maintenance Engineering, Planning and Procurement (SUBMEPP) office; they are thought to be applicable to improving the planning of surface ship maintenance. Appendix B describes the most significant line items in the maintenance plan for *Arleigh Burke*–class destroyers.

CHAPTER TWO

Trends in Surface Ship Maintenance Expenditure: The Record to Date

Introduction

Funding levels for surface ship maintenance are set by the Chief of Naval Operations Director of Surface Warfare (N96), who serves as resource sponsor for surface ship programs. This is based on inputs received from the requirements sponsor (N43) and from requests by the fleets. The type commander (Commander Naval Surface Forces) influences the fleets and also represents interests directly to the resource sponsor. From the start of the process, the resource sponsor is setting priorities among maintenance of the existing surface force and modernization or new construction of the future force.

With funding levels established, the Navy has a rather complicated structure for establishing maintenance and engineering requirements, scheduling availabilities, and funding O&M. Figure 2.1 shows the overlapping and sometimes conflicting chains:[1]

Figure 2.1

Navy Surface Ship Maintenance Organization

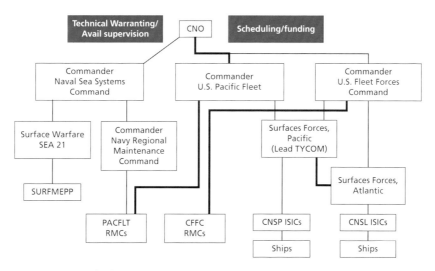

NOTE: PACFLT = Pacific Fleet; RMC = Regional Maintenance Center; CFFC = Commander, U.S. Fleet Forces Command; CNSP = Commander, Naval Surface Forces Pacific Command; ISIC = Immediate Superior in Command; CNSL = Commander, Naval Surface Forces Atlantic
RAND RR1155-2.1

---

[1] Office of the Chief of Naval Operations, *Maintenance Policy For United States Navy Ships*, OPNAV Instruction 4700.7L, Washington, D.C.: Department of the Navy, May 25, 2010.

NAVSEA is responsible for capturing engineering requirements. This specifically means that technical warrant holders[2] state what maintenance actions ships need to reach their ESL. These requirements are captured in TFPs, which are discussed in more detail in later chapters. In addition, NAVSEA is responsible for supervising work carried out in private shipyards, the major providers of maintenance for surface ships. It has, however, no direct ability to reallocate or provide resources.

On the other side of the chain (right side of the diagram), responsibility for scheduling ships for maintenance and providing the funding for availabilities rests with fleet commanders. While NAVSEA can say that failure to correct some conditions makes the ship inoperable, the actual decision to carry out an availability and the specific items completed in availabilities rests with the fleets. The fleets, moreover, must fund both O&M from the same account.

While different resource sponsors provide fleet funding, actual administration of availability planning for every ship class is done through Commander U.S. Fleet Forces Command on the East Coast, Commander U.S. Pacific Fleet on the West Coast, and the Forward Deployed Naval Force in the Pacific. This is provided in a common O&M account, in which fleet commanders are free to move resources between O&M and between ship types.

In practice, O&M has stayed in relatively stable proportion. Figure 2.2, drawn from Navy budget exhibits, shows the relative proportions since 2001. Although there is some variation in proportion, both are increasing and, in fact, maintenance occasionally goes up at a more rapid rate than operations. While there may be opportunity and perhaps incentive to move resources from maintenance to operations, to date this does not seem to have been an issue.

Figure 2.2
Fleet-Wide Operations and Maintenance Expenditures

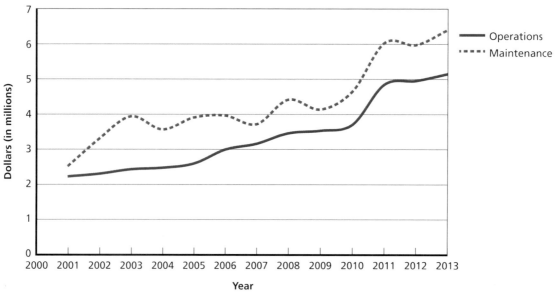

SOURCE: Navy budget exhibits.
RAND RR1155-2.2

---

[2] Technical warrant holders are NAVSEA personnel chosen based on their expertise in a given area and derive their authority from the Secretary of the Navy via the Commander of NAVSEA.

Incentives to perform maintenance, however, on one ship type—or even particular ships—over others do exist. Deployment schedules, emergent repairs, unexpected availability growth, and a variety of other special circumstances can cause one ship to receive more maintenance than expected and another to receive less. That there is impact on the ship not receiving the maintenance might be understood, but the fleet is free to carry these shifts out to support demands of Ao.

Historical Trends

Figure 2.3 below shows the history of surface ship maintenance expenditure. Generally speaking, maintenance cost per ship has increased, although not steadily. The graph shows two dips in maintenance cost per ship. The first dip was in the mid-1980s, when new ships requiring little maintenance were added to the inventory. A second dip was in early 1990s, when whole classes of older, maintenance intensive ships were decommissioned and removed from the inventory.[3] Although there are isolated instances of lower cost, since the early 2000s, the maintenance cost per ship has been rising, even as the numbers in the fleet stabilized. We will discuss this in more detail.

Figure 2.4, drawn from Navy Budget Books, shows that inflation-adjusted maintenance per ship has experienced a strong upward trend since the early 2000s. The blue line shows actual expenditure; and the red line shows an exponential function, which closely matches the overall growth trend. Fleet size has stabilized, so this is not just an artifact of declining numbers. Moreover, such growth is historically unprecedented. The longest and most dramatic period of increase in maintenance cost per ship came following 2004 (Figure 2.4).

Figure 2.3
Surface Fleet Maintenance Expenditures, 1984–2013

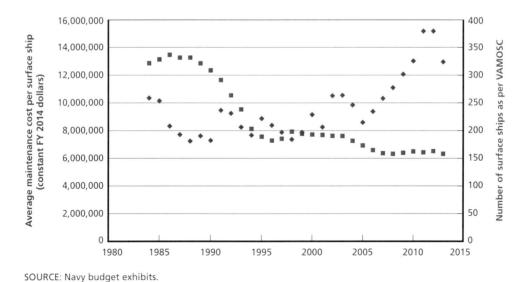

SOURCE: Navy budget exhibits.

RAND RR1155-2.3

---

[3] Charles Nemfakos, former assistant secretary of the Navy, Financial Management and Comptroller, interview with authors, Washington, D.C., June 15, 2014.

Figure 2.4
Navy Surface Ship Maintenance Cost per Ship Trend

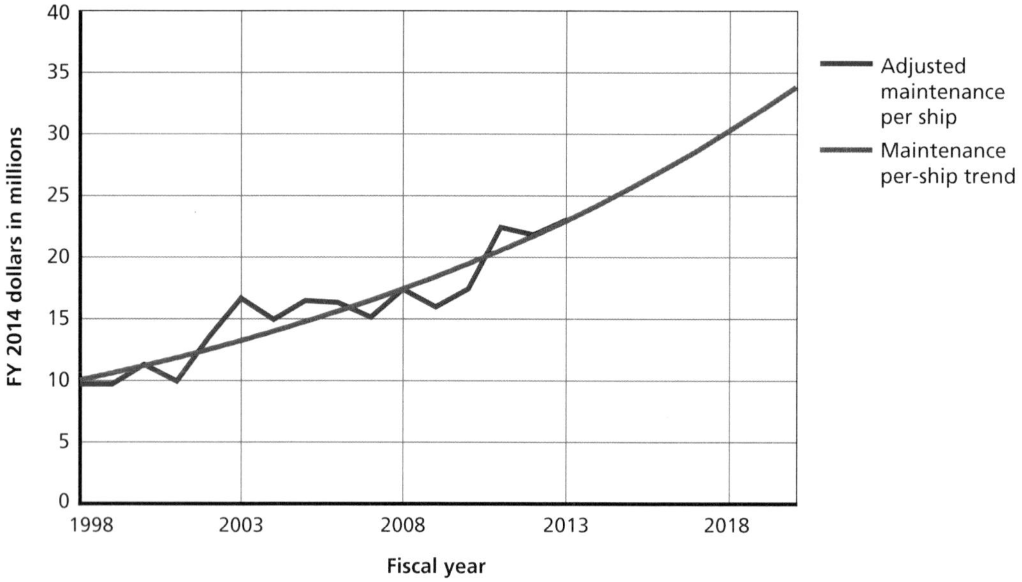

SOURCE: Department of the Navy, "Fiscal Year (FY) 2014 Budget Estimates, Justification of Estimates, Operation and Maintenance, Navy," April 2013.
RAND *RR1155-2.4*

Note that these results are for the entire surface fleet, with results averaged. Bearing in mind that a rising average could be the result of a single ship class or even a small number of ships causing upward pressure, we further subdivided by individual ship class.

The results are shown below. This cost growth is not confined to a particular class of ship. Figures 2.5 through 2.10 depict financial information for DDG-51, CG-47, amphibious assault ship (LHD)-1, LPD-17, LSD-41, and LSD-49 classes on availability execution from the Navy's VAMOSC database. The results vary from class to class, but the story is fundamentally the same: in the last 15 years, expenditure per ship has risen steadily, sometimes dramatically. In a few cases, the reasons for this are clear. For example, LPD-17 suffered a number of casualties early in its life cycle, serious enough to require that the lead ship received multiple pierside availabilities and suffered sufficient materiel failures to warrant a flag-level investigation.[4] In other classes—such as the LSD-49 cargo variant of the LSD-41—the numbers are small, and a single high-profile failure can change the slope of the trend line. The same general trend is seen in the ship class with the most numerous vessels—DDG-51—and appears nearly uniformly across every other class examined.[5] Note that the period in which cost growth was most notable coincided with a period of relative stability in ship numbers, and in fact is seen in ship classes where numbers of ships actually increased over time.

---

[4] Department of the Navy, 2010.

[5] These figures are not presented on the same scale in each graph. Ships are different and can be expected to cost less or more depending on size and complexity. The relevant measure is the degree to which cost growth occurred in each class, not that the average DDG-51 costs less to maintain than the average LHD-1.

Figure 2.5
DDG-51 Maintenance Cost per Ship

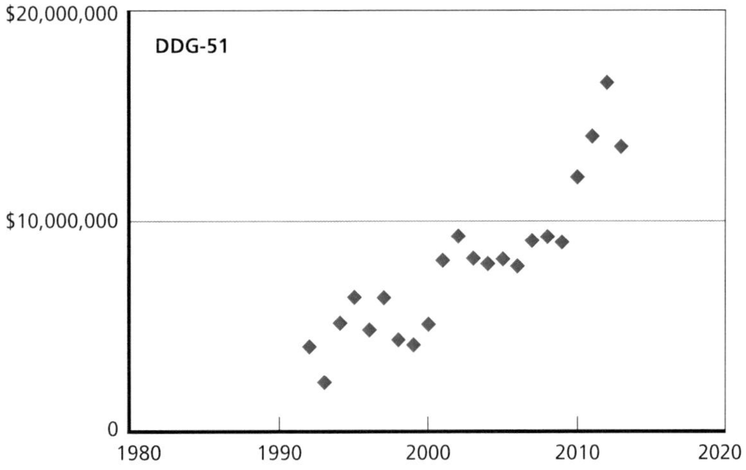

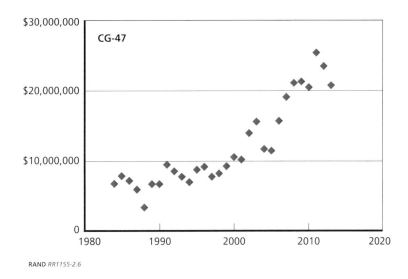

RAND *RR1155-2.5*

Figure 2.6
CG-47 Maintenance Cost per Ship

RAND *RR1155-2.6*

Possible Reasons for Cost-per-Ship Growth: What We Do Not Address

The focus of this report is on the current situation and the potential mechanisms for dealing with it. There are a number of complicated reasons for why growth in maintenance cost per ship have occurred, and not all will be susceptible to alteration or are even relevant to the potential solutions. So while we acknowledge these possible reasons as important and will discuss them briefly, the team and the sponsor determined them to be beyond the scope of the project.

In our review of some possible causes, we determined one to be that the Navy leadership decided to make improved materiel readiness a priority and applied resources to correct it. Interviews with members of the maintenance community suggest that the escalating cost

Figure 2.7
LHD-1 Maintenance Cost per Ship

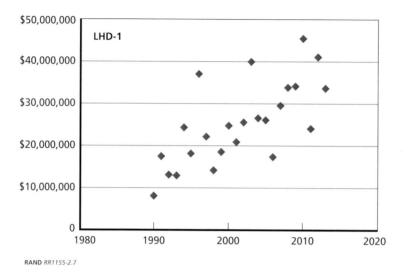

RAND *RR1155-2.7*

Figure 2.8
LPD-17 Maintenance Cost per Ship

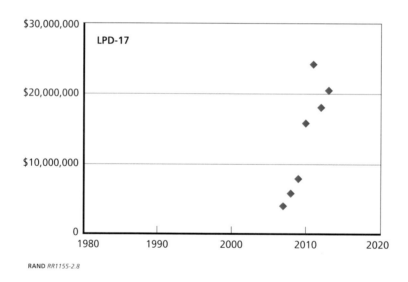

RAND *RR1155-2.8*

might reflect the Navy's desire to place higher priority on ship readiness following the 2010 Balisle report.[6] Indeed, recent public statements by the Commander of Naval Surface Forces may support that view.[7] The Balisle report was issued in 2010 and the observed growth across all ship classes started approximately nine years before then. The report is indeed highly critical of readiness levels during the exact period that maintenance cost per ship was going up most

---

[6]   See Rear Admiral Larry Creevy and Rear Admiral William Galinis, interview with authors, Washington, D.C., August 2014.

[7]   David Larter, "Three-Star: Surface Fleet Readiness, Training Are on Track," *NavyTimes.com*, January 13, 2015.

**Figure 2.9
LSD-41 Maintenance Cost per Ship**

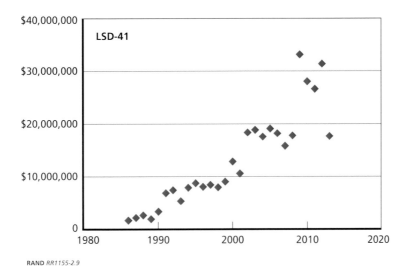

RAND *RR1155-2.9*

**Figure 2.10
LSD-49 Maintenance Cost per Ship**

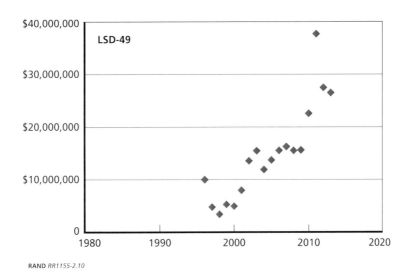

RAND *RR1155-2.10*

rapidly. Factors other than a desire to remedy recently identified shortfalls must have contributed to escalating costs of availabilities.

A related possibility might be that maintenance expenditures per ship grew simply because resources were available, and the Navy saw this as an opportunity to reduce maintenance backlog. Certainly the expenditure of resources is a decision, not an act of nature, and decisionmakers were reacting both to the availability of resources and identified need. As seen in Figure 2.11, however, Navy Total Obligation Authority during the period of rapid growth in cost per ship actually decreased. Something other than simple resource availability must have driven the increased cost per ship.

Figure 2.11
Navy Total Obligation Authority

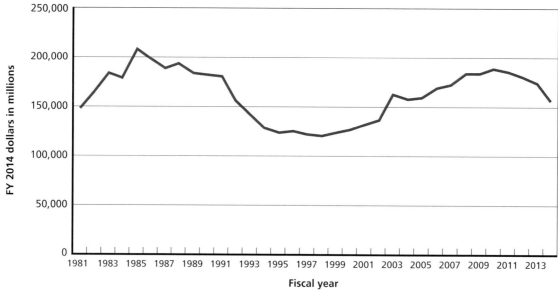

SOURCE: Navy budget exhibits.
RAND *RR1155-2.11*

Another possible cause considered was contract structures and/or shortfalls in shore-maintenance infrastructure. Growth is consistent, however, across ships in every homeport and for a number of different maintenance contractors and in some cases with different contract vehicles. The trends are similar regardless of these considerations.

There may also be diminished crew training for recognition of materiel discrepancies and decreased manning on ships to keep conditions from deteriorating below what the maintenance community expects it will have to correct. Ships may be arriving in availabilities in worse-than-expected materiel condition, thus requiring additional resources and causing maintenance cost per ship to increase. That said, judgments of crew capability to assess conditions are highly subjective and not readily measured. There is no materiel training baseline against which to assess whether ships are better or worse. The upward trend is pervasive across ship classes and time.

Still another reason might be the high operational tempo the smaller force has been required to undertake during a long period of overseas conflicts. We did find no particular association between amount spent on steaming hours and amount spent on maintenance (see Figure 2.12) for at least the DDG-51 class. Steaming hours remained consistent, and ship costs steadily increased. But that does not necessarily prove that there is no relationship between operating for significant periods and as a result seeing more repair requirements. It does suggest that if a relationship does exist, it is not simple.

Getting a complete answer about what might be causing cost growth in most cases will take considerably more research and data than was available to us when we did the work. Both the sponsor and the team concluded that this was out of scope and perhaps not critical to answering the more immediate question of cost containment in the face of constrained resources. We will discuss this in more detail in subsequent chapters.

Figure 2.12
DDG-51 Maintenance Cost and Steaming Hours

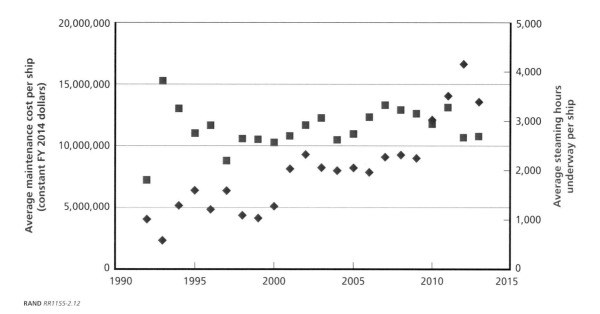

RAND *RR1155-2.12*

Maintenance Trends: What We Do Address

The scope of the study did not include finding the full reason for cost growth. The objective and the scope of the study did, however, aim to find ways for the Navy to contain effectively unprecedented cost growth without undue to impact to Ao and ESL. A major factor in cost growth may simply be that some classes are reaching extensive mid-life overhauls through the period observed. If this is the case, the Navy should have matched resources to predictable maintenance demand. We will examine this in more detail for the DDG-51 class in subsequent chapters.

In addition, the Navy has attempted to document the requirement by ship for particular kinds of maintenance in TFPs published by Naval Sea Systems Command. These are directive in nature and could push cost upwards. We will examine the degree of congruity between the TFP and the actual maintenance expenditure for the DDG-51 class.

Finally, the normal result of curtailment in maintenance activity—due either to time or resources—is deferral of some maintenance actions. When done, it is possible that the condition of the deferred item will worsen, possibly making it more expensive to do at a later time. That may be exacerbated by the normal escalation of labor and materiel cost over time. We will look at both the DDG-51 class and examples from other classes to determine the degree to which this might create cost growth and also to determine ways to effectively manage this.

CHAPTER THREE

The Fiscal Environment

While there are several reasons why the trend toward increased cost might be occurring, it has occurred in the context of generally increasing budgets. Figure 3.1 shows the overall direction of the O&M budget since 1950. Here, we note that maintenance costs have historically been a relatively fixed fraction of O&M. Even in the face of declining fleet size, this budget has been allowed to generally increase over time.

The positive budget environment, however, is very unlikely to persist over the next several years. As alluded to earlier, the BCA of 2011 establishes mandatory spending caps for every agency in government, with built-in mechanisms to ensure that caps are not circumvented as long as the law remains as written. Specifically, even if an appropriation is higher than the BCA cap, the law currently calls for imposition of the lower amount once the compromise Bipartisan Budget Act expires in 2016.[1]

In any case, any of the likely enacted appropriations are below what the Navy has said it needs for O&M. Figure 3.2 shows the difference between what the Navy has said it needs

Figure 3.1
Historical Navy Operations and Maintenance Budgets

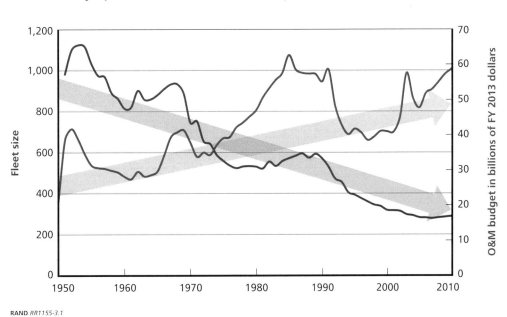

RAND *RR1155-3.1*

---

[1] Douglas W. Elmendorf, letter to John A. Boehner and Harry Reid re Budget Control Act of 2011, Washington, D.C., August 1, 2011; and "Estimated Impact of Automatic Budget Enforcement Procedures Specified in the Budget Control Act," Washington, D.C., September 12, 2011.

Figure 3.2
O&M Funding Scenarios

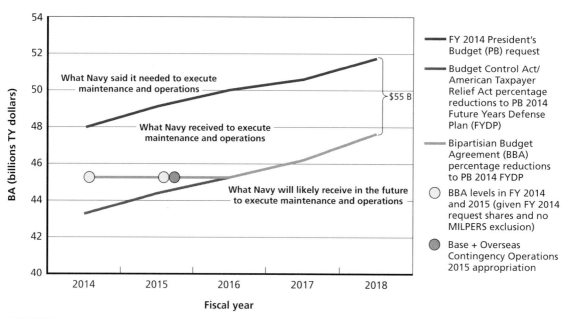

RAND *RR1155-3.2*

to support these areas and what the Navy will receive at best short of complete repeal of the appropriation limiting statutes. Note that O&M share the same account line and are controlled by the same commands—the fleets—and thus compete for resources. If fleets opt to prioritize keeping the fleet underway by attempting to maximize Ao, that will increase the pressure maintenance resources.

The fundamental point is that the fiscal environment points to constraint and that the mechanisms for dealing with constraint generally involve trade-offs with other priorities. The Navy cannot count on expanding resources, and it should also be prepared to deal with resources well below what it has requested as necessary.

This runs squarely into the previously noted trends in escalating per-ship maintenance costs. Figure 3.3 puts a BCA-constrained line into Figure 2.4 on cost-per-surface-ship maintenance trends adjusted to FY 2014 dollars. The difference between where the trend is taking the Navy and where the budget would allow is significant, in the tens of millions of dollars per ship. If both trends follow this projected path, this would result in a significant mismatch between requirement and resource.

In the case of DDG-51, the largest surface ship class in the Navy, there is good reason to expect unconstrained maintenance costs increases to continue. To explain, mid-life maintenance, known as *Extended Docking Selected Restricted Availability* (EDSRA), is the most expensive maintenance availability in the service life of DDG-51 destroyers. An EDSRA is a Docking Selected Restricted Availability (DSRA) expanded to include maintenance and modernization that cannot be accomplished in a regular DSRA. With EDSRA, notionally scheduled to begin at 192 months (16 years), 17 DDG-51 destroyers have completed their EDSRA in the past six years; destroyers having gone through EDSRA are now 19 to 24 years old. As shown in Figure 3.4, 26 more destroyers will have to go through EDSRA in the next five years to meet the goal of EDSRA at 16 years. Put another way, the pace of EDSRAs today has been just under

Figure 3.3
Navy Surface Ship Maintenance Trends

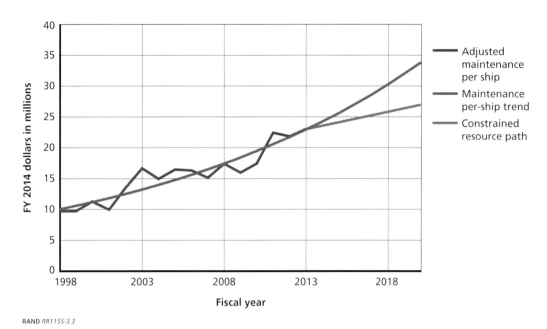

RAND RR1155-3.3

Figure 3.4
The Goal of Extended Docking Selected Restricted Availability at 16 Years Is Slipping

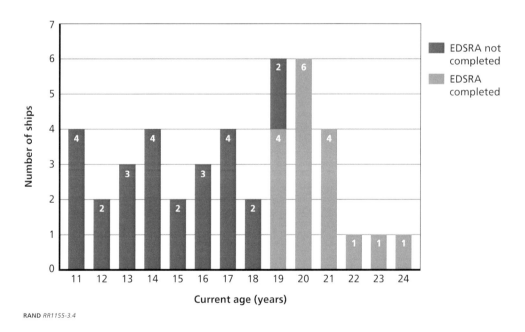

RAND RR1155-3.4

three (17 divided by 6) EDSRAs per year. Continuing at this pace will risk further deferring maintenance scheduled for EDSRAs; at the current pace, the EDSRAs of DDG-51s now 11 years old will be deferred by four years until they are 20 years old. We will discuss in future chapters some potential consequences of extensive deferral as measured in costs of availabilities following extensive deferral, but for this discussion, it is safe to say that deferral results in compounded deterioration in materiel condition. Moreover, as will be discussed later, the maintenance availabilities scheduled after EDSRA are expected to be more expensive than the maintenance availabilities scheduled before EDSRA. Lead ships of the DDG-51 class are now moving past mid-life and into this period of increasingly expensive maintenance.

Processes to Support Fleet Materiel Readiness

The Navy is aware of surface ship maintenance issues and has attempted to implement processes to improve or at least foster improvement in ship-maintenance planning and execution. These do not directly address the issue of cost-per-ship maintenance escalation, but they may influence planning and resource allocation processes that might. The Optimized Fleet Response Plan (OFRP) is intended to promote better planning and standardization of maintenance and modernization packages.[2] Although this also applies to training and personnel management, it applies at present only to ships assigned to Carrier Strike Groups (CSGs) and to all elements within the CSG. The OFRP creates standard maintenance periods and ensures that these are maintained even in the face of other demands. By establishing fixed training periods and aligning ships consistently to strike groups, the OFRP also encourages type commanders (TYCOMs) and NAVSEA to plan work packages well in advance and to ensure adequate time and resources. Still, with all that said, OFRP might provide incentives, but it does not bind the maintenance community to search for lower cost or more effective delivery of service.

Surface Maintenance Engineering Planning Program (SURFMEPP) was implemented in 2010 to identify and defend maintenance requirements.[3] This program is based on TFPs, but is informed by identification of ship-specific conditions. This promises to eventually be an effective tool for understanding what maintenance has been accomplished and what might still need to be done. While SURFMEPP tracks the accomplishment of ship-specific requirements, it does not establish the requirements, which is done by technical warrant holders. SURFMEPP may challenge requirements based on observed data, but it neither sets nor reduces requirements and can in effect only recommend funding and execution decisions be made by type and fleet commanders. There is no guarantee that this will improve or even affect their decisionmaking. It is also hampered by being a new process for which critical data are still being collected. While it holds long-term promise, it has not yet fully matured as an organization and needs additional time and data collection before it fulfills this promise.

---

[2] The OFRP replaces the previous Enhanced Carrier Presence concept and reorganizes processes for training, inspections, parts, maintenance, and manning to increase efficiency, stability, and predictability across the fleet. It increases homeport tempo from 49 percent to 68 percent over 36-month periods.

[3] SURFMEPP is an echelon-three command of the NAVSEA comprising approximately 240 (at end strength) personnel, including military (Commanding Officer), Civil Service personnel, and contractor technical-support staff. Headquartered at Norfolk Naval Shipyard in Portsmouth, Virginia, SURFMEPP maintains detachments located in Norfolk; Mayport, Florida; Pearl Harbor, Hawaii; Yokosuka, Japan; Sasebo, Japan; Everett, Washington; San Diego, California; Manama, Bahrain; and Rota, Spain. The Navy established the SURFMEPP on November 8, 2010, to provide centralized surface ship life-cycle maintenance engineering, class maintenance and modernization planning, and management of maintenance strategies aligned with and responsive to TYCOM and NAVSEA needs and priorities.

Concluding Observations

The findings in this chapter need to be kept in the overall resource context and be considered solely in context of per-ship cost growth. The Navy will likely not decommission big blocks of old ships and replace them with new. There simply are not enough new hulls being considered to meet anything like the demand requirement. Realistic options will have to rely heavily on maintaining (and likely modernizing) older platforms, which we can see are experiencing steady escalation in the cost it takes to maintain them. If ship maintenance continues on the increased cost-per-ship track and retains its normal portion of the O&M account, however, significant mismatches will occur between available resources and need.

We will discuss more fully some potential mitigations for such a shortfall, but it will be important for the Navy to recognize that the cost-escalation trend and the likely budget trends are moving in unhealthy directions.

Current Maintenance Strategies

This chapter first describes the existing surface ship maintenance strategy, then introduces two potential new surface ship maintenance strategies intended to minimize negative effects on ESL and Ao in order to maintain the largest, most capable fleet possible. The alternative strategies are offered in the context of an era of limited time and money for maintenance and in which risks must be accepted. Current and potential new strategies are illustrated for *Arleigh Burke* class (DDG-51), but are intended to apply more broadly.

We first provide an overview of the process, then describe perceived potential areas in which it could be improved. Two potentially improved surface ship maintenance strategies are then described.

The Current Maintenance Strategy

OPNAV Instruction 4700.7L, "Maintenance Policy For United States Navy Ships," dated May 25, 2010 (Office of the Chief of Naval Operations, 2010), states the goal of conducting surface ship maintenance only on actual failure of a ship's installed systems or components, or it is possible to predict failure with confidence. The Naval Sea Systems Command (COMNAVSEASYSCOM)–approved Reliability-Centered Maintenance (RCM) methodology is the prescribed means of predicting failures.

The RCM methodology originated in the airline industry in the 1950s. The underlying rationale then and now for preventive maintenance planning is the belief that the reliability of hardware decreases with age and use. The Navy first used the RCM methodology in 1978 using FF-1052 class frigates. The Phased Maintenance Program, developed in the early 1980s, expanded the application of the RCM methodology to ship depot–level maintenance. This program was subsequently expanded to the continuous maintenance program, and later to conditions-based maintenance, which requires the use of RCM to determine evidence needed to select appropriate maintenance.

NAVSEA's *Reliability-Centered Maintenance (RCM) Handbook* states that, after initial success, benefits derived from the use of classic RCM to establish maintenance programs eroded over time due to failure to conduct periodic analysis of O&M feedback to improve the periodicity and scope of prescribed maintenance tasks. COMNAVSEASYSCOM developed an improved RCM methodology to correct this situation in 1996. The application of the improved methodology (the so-called RCM backfit methodology) reduced the Planned Maintenance System workload on ship's force by nearly 50 percent without adversely affecting safety, mission, or environment. The results of this process evolution are the classic RCM approach and the RCM backfit methodology for validating maintenance requirements for in-service systems and equipment.[1] The SURFMEPP within NAVSEA implements depot-level maintenance planning through class-specific TFPs.[2] We reviewed the TFPs for *Arleigh Burke* (DDG-51) and *San Antonio* (LPD-17) classes. Their position and functions within programming and execution cycles are described before describing and discussing those TFPs.

Surface Ship Maintenance Cycles

The position of TFPs in the surface ship maintenance programming and execution cycles is illustrated in Figure 4.1, which is taken from the DDG-51 TFP. Two cycles, "Programming and Execution," and the interaction between the two cycles are shown.

Figure 4.1
Position of TFPs in Surface Ship Maintenance Process

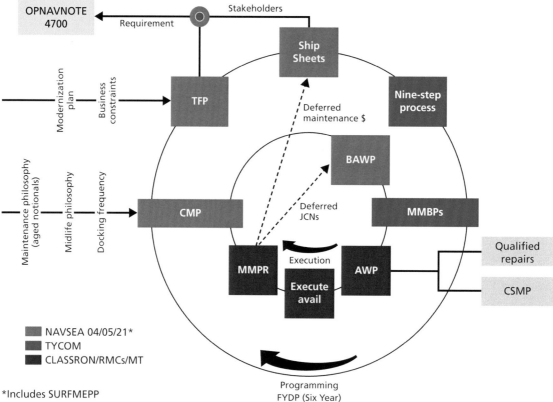

NOTE: AWP = availability work package; BAWP = Baseline Availability Work Package; CSMP = Current Ship Maintenance Project; JCN = Job Classification Number; MMBP = Maintenance and Modernization Business Plan; MMPR = Maintenance and Modernization Performance Review; OPNAVNOTE = Office of the Chief of Naval Operations Note; RMC = Regional Maintenance Center

RAND *RR1155-4.1*

---

[1] In more detail, classic RCM is a methodology that carefully develops, analyzes, and documents requirements thoroughly as it proceeds to develop a maintenance program in an environment of uncertainty with limited operating data. Backfit RCM applies where sufficient operating data exists to confirm assumptions that were made when the original maintenance program was developed, validating existing maintenance requirements, or recommending changes as appropriate. Classic RCM provides a technique for dealing with the uncertainty of maintaining new equipment for which no operating history exists. Backfit RCM provides a technique for dealing with the uncertainty of changing existing, satisfactory maintenance requirements to make the maintenance program even more effective.

[2] Office of the Chief of Naval Operations, 2010.

Relevant components of the overall process illustrated in Figure 4.1 are described below beginning with the Class Maintenance Plan (CMP) and continuing through other steps in the cycle.

Class Maintenance Plan

The CMP, a maintenance manual for each ship class, is considered to be a core product of SURFMEPP. It consists of the maintenance and assessment tasks required for ship class to achieve its ESL. CMPs are updated based on technical analyses and periodic reviews with associated technical authorities.

CMPs include the following:

- Maintenance delivery plan, which includes dry-docking intervals
- Engineered maintenance requirements, such as equipment overhauls, shaft replacements, and corrosion protection
- Special system certification imposed during surface ship industrial periods to systems, such as boilers and operational software.

Technical Foundation Papers

TFPs serve two functions within surface ship maintenance processes. First, the TFP is a programming tool; it is the vehicle by which Office of the Chief of Naval Operations Note (OPNAVNOTE) 4700 requirements are revised; each TFP documents the outcome of a full technical analysis of the requirements over the life cycle of the class. Second, the TFP is an execution-tool intermediate between the CMP and Ship Sheets (more in the next section). TFPs for the DDG-51 and LPD-17 classes are described and discussed later in this chapter.

Ship Sheets

Ship Sheets document labor and materiel costs associated with each ship's Chief of Naval Operations (CNO) availability maintenance requirements in support of the Program Objective Memorandum (POM) process and support ships meeting their ESL.[3] Ship Sheets are a critical tool in the POM budget process; they are used to clearly articulate the resources required to perform the needed ship maintenance.

In addition to supporting the POM process and helping ships meet their ESLs, Ship Sheets describe planned and previously deferred maintenance requirements. They are developed by comparing the current condition of the ship against the TFP to determine specific maintenance tasks required to fix the ship.

Nine-Step Process

The Nine-Step Process is used to rebalance resources across the force-maintenance schedule for each FY within the limits of OPNAVNOTE 4700 requirements. The process is based on availability information provided by Ship Sheets. The Nine-Step Process provides a hull-by-hull review of individual ship-maintenance requirements to better refine notional ship main-

---

[3] There are three basic types of maintenance availabilities: CNO availabilities, non-CNO (TYCOM) availabilities, and new construction (SCN) availabilities. CNO availabilities are relatively long and are scheduled with CNO approval. CNO availabilities include SRAs, which are discussed later. Non-CNO availabilities include Continuous Maintenance Availability (CMAV) periods, which are also later discussed. SCN availabilities include post-shakedown availabilities. SCN availabilities are not discussed here.

tenance requirements and tailor them to the physical condition of individual platforms as they approach the time when they will be inducted into their scheduled availability periods. The process includes expected shipyard performance, shipyard capacity considerations, and develops alternative courses of action for the completion of any work requirements that would exceed the available capacity.

Maintenance and Modernization Business Plan

Maintenance and Modernization Business Plans (MMBPs) are annual maintenance budgets for each ship of a class in a given FY. Each MMBP is developed by the applicable regional maintenance center maintenance team and is issued by the TYCOM.

Baseline Availability Work Packages

Baseline Availability Work Packages (BAWPs) contain CMP tasks as specified by the TFP; deferred CMP tasks from the previous cycle as identified in Ship Sheets and specified in the MMBP; work identified by CMP assessments deferred from the previous cycle; and temporary services, known modernization, and TYCOM mandatory programs (habitability renewal). BAWPs are unique to each ship, covering life-cycle maintenance requirements, including continuous maintenance from the completion of one CNO availability to the completion of the next CNO availability. BAWPs are developed approximately 22 months prior to the start of the next CNO availability.

Availability Work Packages

Availability Work Packages (AWPs) are one of two bridges between the execution and programming cycles. Inputs to AWPs come from BAWPs and MMBPs; they combine BAWP-identified tasks with the depot-level corrective maintenance from the Current Ship's Maintenance Project (CSMP) as well as additional work identified by CMP assessments after the BAWP was issued.

Availability Execution, Close Out, and Maintenance and Modernization Performance Review

SURFMEPP issues a BAWP closeout report following the execution and completion of an availability. The report details each BAWP item, its resolution, and the applicable cost data. The closeout report also records deferred work. This information is an input to the Ship Sheets and the BAWP for the next cycle. The report will also include feedback on CMP tasks (input to the CMP) and can be used to compare the notional requirement with the work actually executed (input to future TFP revisions).

The Maintenance and Modernization Performance Review (MMPR) is a process review of the CNO availability held within two to four weeks following the completion of the availability. The objective of the MMPR process is to identify process improvements as they pertain to the execution and planning of the availability (how the work gets done); the BAWP closeout process is focused on improving the accuracy of the maintenance requirement (what work gets done).

Continuous-Maintenance Availabilities

In addition to maintenance conducted within ship maintenance cycles, ship maintenance is conducted pierside within Continuous Maintenance Availability (CMAV) periods. CMAVs are the only type of availability accomplished on ships outside of CNO availabilities for non-emergent maintenance. CMAVs can be scheduled or unscheduled; scheduled CMAVs normally occur between Selected Restricted Availabilities but can be assigned on a concurrent basis with CNO-scheduled depot availabilities. CMAVs normally last two to six weeks (but can extend to more than six months[4]). They are normally scheduled when the ship is in port.[5] CMAVs are intended for accomplishment of inspections, condition-based upkeep, and minor repairs. The inspection process is driven by the CMP and is a foundation for CNO availability work package development. OPNAVNOTE 4700 (August 11, 2011) budgets 1,600 man-days (continental United States [CONUS]) and 2,100 man-days (forward deployed naval forces [FDNF]) per year for CMAVs. The DDG-51 TFP recommends a notional CMAV man-day requirement of 1,700 man-days (CONUS Flight [Flt] I/II), 1,900 man-days (CONUS Flt IIA), and 3,500 man-days (FDNF Flt IIA). CMAV requirements were developed in conjunction with the CNO requirements; they do not overlap.

By design, CMAVs provide maintenance teams with flexibility. As a result, CMAVs are not as well defined as CNO-scheduled maintenance availabilities. Some complex CMAV activities (such as hull, machinery, electrical, electronics, piping work, and ship alterations) are contracted out competitively. Other CMAV activities (such as painting or inspections) can be conducted by ships' company. CMAVs therefore represent a flexible and economic means to accomplish inspections, maintenance, and repairs pierside. CMAVs can be more cost effective than other CNO availabilities. Reasons for this include reduced administration and security costs and reduced energy costs.

DDG-51 TFP

Increasing surface ship Ao by optimizing advanced planning and execution of depot-level maintenance is one stated goal of the TFP for *Arleigh Burke*–class destroyers. Early identification of maintenance tasks is said to result in more accurately planned maintenance periods, minimized cost and schedule growth during maintenance execution, and stable maintenance budgets. The extent to which the maintenance community is able to follow the guidelines in the TFP is also said to be a factor determining actual ESL.

The TFP for *Arleigh Burke*–class destroyers is stated to have been developed on the basis of an extensive evaluation of current technical requirements and an in-depth statistical analysis of previously executed depot availabilities. It is stated to be based on a Long-Range Maintenance Schedule (LMRS) that was developed in conjunction with the CMP and historical cost trends. The LMRS itself is stated to have been developed by SURFMEPP using technical requirements and historical data from the following sources:[6]

- previous (2009) TFP for the *Arleigh Burke* class
- ships maintenance materiel management system data
- the corrosion control information management system (CCIMS) program
- Navy ships
- technical manual requirements
- Navy maintenance database

---

[4] For example, the USS *Wasp* (LHD-1) conducted an eight-month CMAV in 2010 and 2011.

[5] Office of the Chief of Naval Operations, 2010.

[6] This study was unable to access these sources.

- departure and availability status reports
- CSMP
- the master assessment index
- board of inspection and survey data
- top management attention/top management issues
- inputs from TYCOM N43 type desk officers/assistants and maintenance teams
- the Naval Surface Warfare Center Dahlgren Division
- departure from specification reports
- and the Maintenance Resource System (MRS) database.

Depot-level maintenance plans for *Arleigh Burke*–class ships are prescribed by flight; there are plans for Flight I/II and Flight IIA ships. FDNF have availability schedules reflecting their higher operational tempo. All availabilities are currently scheduled at 32-month intervals and are predicated on expected 35-year ship service lives.

The *Arleigh Burke*–class life-cycle maintenance plan consists of three parts: (1) technically mandated assessment and depot-maintenance requirements, (2) notional estimates of man-days sufficient to execute those requirements, and (3) scheduled maintenance periods to complete those requirements. Maintenance is keyed to Selected Restricted Availability (SRA), DSRA, and EDSRA periods. The LMRS assigns man-day estimates across the life cycle in notional man-days. These estimates are organized by Ships Work Line Item Numbers (SWLINs) to estimate the total cost of maintaining Flight I/II ships across an ESL of 35 years. All man-day estimates are stated to be based on maintenance community inputs. Aging factors have been applied to man-day estimates across the full life cycle.

SWLINs are organized in the following series:

- Series 100, hull structure
- Series 200, propulsion plant
- Series 300, electric plant
- Series 400, command and surveillance
- Series 500, auxiliary systems
- Series 600, outfit and furnishings
- Series 700, armament
- Series 800, ship assembly and support and support services

The LMRS currently has a total of 2,368 SWLINs. Forty-four SWLINs, or groups of SWLINs, are identified in the DDG-51 TFP as having significant life cycle effect or significant cost drivers. They are summarized in Appendix B.

Flight I/II cumulative man-days per hull by ship age, as shown in the DDG-51 TFP, are presented in Figure 4.2.

The figure indicates that beyond the age of eight years (DSRA-1), all but one DDG-51 destroyer received less maintenance than is required in the DDG-51 TFP; the exception appears to be the USS *Cole* (DDG-67), which was severely damaged in a terrorist attack when she was five years old and underwent extensive repairs. Equally important, there appears to be little or no effort to close the gap between required and actual maintenance prior to the mid-life EDSRA.

Figure 4.2
DDG-51 Flight I/II Cumulative Man-Days per Hull by Age

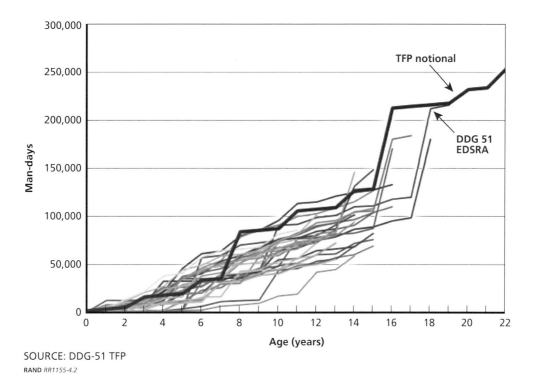

SOURCE: DDG-51 TFP
RAND *RR1155-4.2*

LPD-17 TFP

The TFP for the LPD-17 class is far less developed than the TFP for DDG-51 destroyers due to the relatively young age of the class. It states:

> Data from Maintenance Resource System (MRS) and Ships Maintenance Material Management System (3M) was insufficient due to the young age of the Class. Considering this historical data deficiency, a parametric comparison of systems, major components, and Ships Work Line Item Number (SWLIN) groups was conducted. Thorough analysis of requirements execution during specific LPD 17 Class availabilities will be pursued as part of the TFP revision process.[7]

Worse, issues described by the Navy's inspector general as "shoddy construction and basic workmanship problems" have driven high maintenance levels early in the service lives of LPD-17 class ships. Results analogous to those shown in Figure 4.2 would be meaningless. LPD-17 maintenance issues are not discussed further in this report.

---

[7] Naval Sea Systems Command, "Technical Foundation Paper for LPD 17 Class Revised Type Commander Notionals," Washington, D.C.: U.S. Navy, 2011, p. 5.

Operational Availability

OPNAV Instruction (OPNAVINST) 3000.12A,[8] updates the Navy's policy regarding Ao as a primary measure of readiness of naval systems, subsystems, and equipment, and provides definitions and equations for calculating Ao. This instruction points out that, while mean time between failure is a critical term in measuring Ao, it has limited value in making maintenance decisions. OPNAVINST 3000.12A instead suggests using a random failure model in making certain maintenance decisions. In particular, the publication differentiates between Ao calculations for electronic systems and for Hull, Mechanical, and Electrical (HM&E) systems. Electronic components are subject to random failures; their failures only can be described as the probability of failure over a period of time. Failures for HM&E systems can be predicted based on observations and wear-and-tear experience. For example, a bearing will have a historical wear-out rate based on values for temperature, pressure, and operating time. Use of this concept allows development of preventive maintenance schedules based on predictable failures.

The Naval Sea Systems Command activity SUBMEPP began generating age and reliability curves using maintenance feedback data in 1998. These curves provided the then-young organization with an objective means to measure the effects of planned maintenance to engineer optimal maintenance plans. The resulting processes were originally applied to *Los Angeles*–class submarines (SSN-688) and remain in use today.[9] Like SURFMEPP, SUBMEPP uses principles from RCM. In doing so, SUBMEPP confirmed the two distinct ways identified in OPNAVINST 3000.12A in which hardware can fail: hardware may wear out (resulting in predictable failures) or may have random (unpredictable) failures.

The value of failure-rate analysis is reflected in a case study conducted by SUBMEPP on *Trident*-class ship, submersible, ballistic, nuclear torpedo tubes. Torpedo tubes are critical to the operation of all modern submarines, and the original maintenance plan for them included time-based maintenance to replace hydraulic cylinders every 160 months to prevent external leakage of hydraulic fluid. The case study revealed that the failure pattern for the hydraulic cylinders was random over time; no evidence supported their replacement at 160 months. Instead, procedures to detect degradation in their condition were developed and implemented. Submarine-maintenance requirements were reduced, and millions of dollars in costs avoided.

The net value of maintenance is an important concept in the SUBMEPP process. At some point, the majority of items that are going to fail have failed, so maintenance provides little additional reliability. Dramatic failure-rate decreases may be followed shortly by a swift return to the status quo. The benefits of maintenance in these cases are transitory, and the maintenance itself is not cost effective. Appendix A discusses this in more detail.

Takeaways from the discussion above include that maintenance decisions should consider failure patterns (random or wear out), the duration of benefits, net improvements, and costs compared to benefits.

---

[8] Office of the Chief of Naval Operations, "Operational Availability Handbook: A Practical Guide for Military Systems, Sub-Systems and Equipment," enclosure in *Operational Availability of Equipments and Weapons Systems*, OPNAVINST 3000.12a, N40, Washington, D.C.: Department of the Navy, September 2, 2003.

[9] Timothy M. Allen, *U.S. Navy Analysis of Submarine Maintenance Data and the Development of Age and Reliability Profiles*, Portsmouth, N.H.: Department of the Navy, 2001.

Cost of Deferred Maintenance

Among the factors driving a determination of whether to defer maintenance is the cost of failing to perform maintenance on an item. The direct cost will vary by item, ranging from the case where not performing maintenance results in failure to ones in which the maintenance really has no relationship to the failure or reliability of the component. Most cases fall somewhere between the extremes and would require more extensive failure-rate analysis than we perform here. This may be an additional value of SURFMEPP as it accumulates data and trends.

For our purposes, we look at whether there is evidence of cost growth resulting from periods in which there has been heavy use of a platform, followed by extensive availabilities. We first examine the DDG-51 class, in keeping with the analysis we have performed to date, then look at two older classes that may have had more opportunity for deferring maintenance: LSD 41/49 and two ships from the CG-47 class. This may help up characterize the likely effect of future deferrals on ships currently in service.

Deferred Maintenance for *Arleigh Burke*–Class Destroyers

Any gap between actual ship maintenance and maintenance prescribed in the DDG-51 TFP is considered here to represent deferred maintenance. Looking at the first eight years of ship life, no clear pattern of deferred maintenance is seen (Figure 4.3). After eight years only one DDG-51 (perhaps the USS *Cole*) is seen to have higher maintenance levels than those prescribed in the DDG-51 TFP.

Figure 4.3
DDG-51 Flight I/II Cumulative Man-Days per Hull by Age

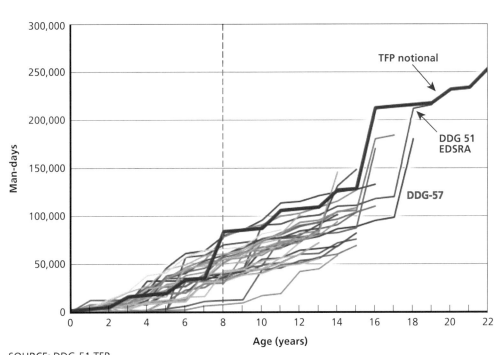

SOURCE: DDG-51 TFP.
RAND *RR1155-4.3*

Man-day rates can be estimated for DDG-51 maintenance using the data portrayed in Figure 4.2, which shows cumulative class-wide man-days. Within the class, we focus on the USS *Mitscher* (DDG-57), a Flight I *Arleigh Burke*–class destroyer. DDG-57 was selected in part because nothing in her history indicated an event (such as a collision) that would skew her maintenance history. The USS *Mitscher* is typical both in terms of cumulative man-days of maintenance and her maintenance schedule. Commissioned in 1994, *Mitscher* entered her EDSRA in February 2013 and completed her EDSRA in November of that year. As shown in Figure 4.3, DDG-57 has a maintenance history typical of Flight I/II *Arleigh Burke*–class destroyers. As shown in Figure 4.4, a nominal man-day rate of $500 per day aligns the results.

An examination of the maintenance histories of all *Arleigh Burke*–class ships, as exemplified by DDG-57, indicates that their cumulative maintenance levels are steadily falling behind the TFP-recommended levels. To date, this has not resulted in cost escalation in out-year availability packages, i.e., the fact that maintenance is falling behind has not resulted in greater expense later in the ship's life. In addition, we could find no obvious evidence that DDG-51s were missing deployments or experiencing materiel degradation during deployments. This could either reflect an overstated requirement—some of the deferred maintenance was not critical—or it may reflect the relative newness and size of the class. Older ships may in fact begin to experience the phenomenon, but it may not be reflected in average class maintenance cost. If this is the case, the deferred actions may in fact pile up significantly and result in even greater upward pressure on ship-maintenance cost. To evaluate the potential effect, we looked at other ship classes with a longer maintenance history and with more obvious potential for deferral.

Figure 4.4
DDG-51 Flight I/II and DDG-57 Cumulative Cost per Hull by Age

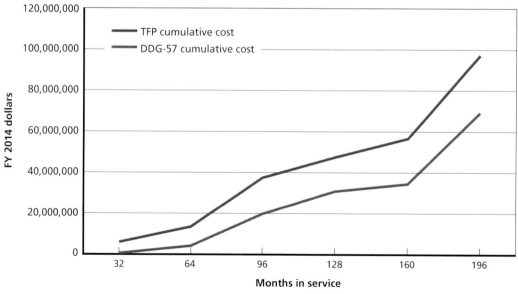

Maintenance Deferred in Other Classes

We looked at two other classes to see whether deferral might be having an effect. In doing this, we did not look at operational history, but strictly at maintenance financials from VAMOSC. We also specifically looked at ships that had been stationed in FDNF, where the maintenance operations construct specifically accepts that some long maintenance periods might not be feasible given the overall pace of operations.[10]

Figure 4.5 shows the maintenance history of the USS *Germantown* (LSD-42), which was present in FDNF from 1993 to 2002.

This particular ship shows both a sizeable availability following its period in FDNF, followed closely by the LSD mid-life availability. To compare this ship to the whole ship class, the ships were divided into three groups:

- LSD-42 and LSD-43, which were in FDNF and have post-FDNF maintenance histories sufficient to determine the effects of prolonged operation with limited maintenance.
- LSD-41, LSD-44, LSD-45, LSD-47, and LSD-48, which were never in FDNF or recently entered FDNF. These ships form the basis for comparison with LSD-42 and LSD-43.
- LSD-46, which recently completed her FDNF deployment. The effects of prolonged operation with limited maintenance cannot be determined for this ship.

The consequences to CNO availabilities subsequent to FDNF periods are apparent in Figure 4.6, with maintenance costs spiking in the first Selected Restricted Availability (SRA) following FDNF periods.

Figure 4.5
Maintenance History of USS *Germantown*

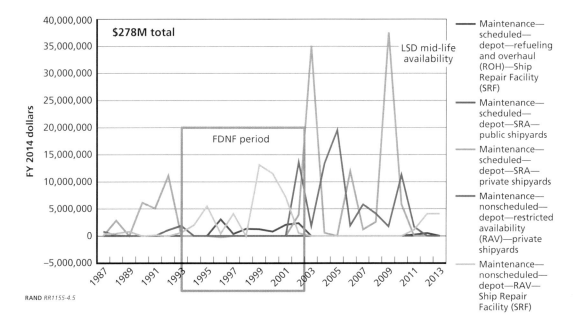

[10] Office of the Chief of Naval Operations, *Optimized Fleet Response Plan*, OPNAVINST 3000.15, USFF/CNO N3/N5, Washington, D.C.: Department of the Navy, November 10, 2014.

Figure 4.6
LSD 42/43 Had High Post-FDNF Maintenance Costs

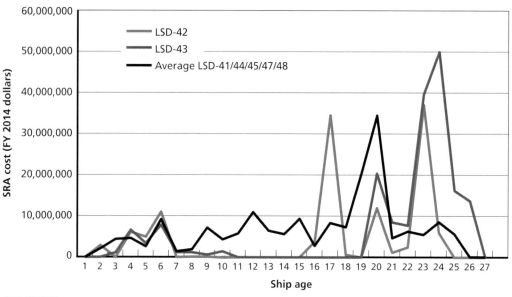

RAND *RR1155-4.6*

A similar pattern in maintenance costs is seen in a comparison of maintenance costs for the cruisers USS *Monterey* (CG-61) and USS *Chancellorsville* (CG-62). These two cruisers, both MK 41 Vertical Launch System variants, were commissioned about six months apart and have comparable operational histories other than that CG-62 was in FDNF. As was the case with LSD-42 and LSD-43, CG-62 experienced spikes in maintenance costs following her return from FDNF (Figure 4.7). More compellingly, the total cost of maintenance for CG-62 ($169 million) has been slightly over twice that of CG-61 ($84 million), as shown in Figure 4.8. No reason for this cost doubling other than stationing is obvious.

Costs of Deferral

The history of the DDG-51 class—the most numerous surface ship class—is still being written. It is clear, however, that there is a mismatch between what the TFP would have prescribed for maintenance and what was actually performed. Whether this will result in significant out-year growth remains to be seen. A look at other ship classes, however, does suggest that operating ships for extended periods and deferring maintenance during those periods have resulted in cumulative maintenance costs in excess of ships experiencing more regular maintenance. Operation in FDNF is one situation where deferrals might occur, but the suggestion is that any situation where maintenance is deferred to support increased operations might cause this kind of pressure. SURFMEPP's ability to individually track deferred-maintenance items should improve our understanding of the specific impacts of deferrals on systems and ships.

These costs will become important as we consider the impact of reduced budgets. When fewer resources are available, there will be continuing pressure to reduce expenses in the near term, and scheduled maintenance is typically a tempting target. While this may reduce pressure in the short term, evidence suggests that, in the longer term, this decision may add to the exponential pressure we already see in maintenance cost per ship.

Figure 4.7
CG-62 Repair Work Following FDNF

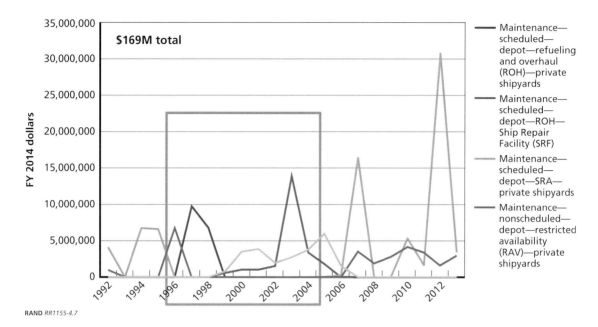

RAND *RR1155-4.7*

Figure 4.8
Maintenance History of CG-61

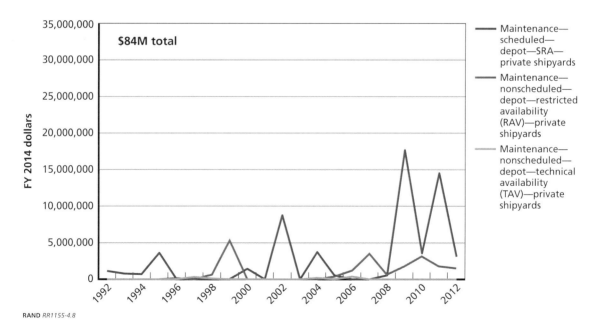

RAND *RR1155-4.8*

CHAPTER FIVE

Potential New Maintenance Strategies

SURFMEPP has developed a comprehensive process for achieving Ao and ESL for surface ships. The basis of this process, as discussed in the previous chapter, is the CMP, TFPs, Ship Sheets, and BAWPs. These are used to plan individual ship availabilities. Availabilities are organized as CNO depot availabilities, CMAV, and CM windows of opportunity. For CONUS-based DDG-51s, the large majority (91 percent) of maintenance man-days assigned to maintenance plans goes to CNO availabilities; for FDNF DDG-51s, all maintenance man-days are assigned to CNO availabilities.

The DDG-51 TFP is detailed and extensive in prescribing maintenance requirements. As demonstrated earlier, actual maintenance accomplished for DDG-51 class ships has been below the TFP notional requirements as a result of cost and schedule limitations, which are only expected to worsen over coming decades. Forty-four SWLINs or groups of SWLINs (presented in Appendix B) are identified in the DDG-51 TFP as having significant life-cycle effect, significant cost drivers or both. The DDG-51 TFP does not recognize cost and schedule limitations in implementing prescribed maintenance requirements. As a result, its treatment of cost focuses on identifying the highest-cost maintenance drivers and does not support cost-benefit decisions. As shown in Appendix B, the most significant SWLINs are characterized as high-cost drivers or life-cycle critical, or as necessary to address and execute to achieve ESL. Duration of benefits and net benefits are not considered. Consequently, there is little basis for selecting maintenance most critical to achieving ESL. Risk is treated in a single sentence, stating that deviation from the TFP's maintenance schedule introduces a life-cycle maintenance risk that could potentially affect the ship's ability to execute its mission in the near term and achieve ESL in the future. In particular, the characteristics of risk are not addressed. A far more extensive and nuanced treatment of risks and risk factors is needed. The option of reallocating maintenance from CNO availabilities to CMAVs is not recognized.

Two potential new maintenance strategies are introduced here: one for prioritizing maintenance within SRAs (recognizing that not all required SWLINs may be executable) and a second for reallocating maintenance from SRAs to CMAVs (possibly with the objective of lowering costs and certainly with the objective of increasing Ao while reaching full ESL). Both strategies are described using decision trees. The purpose of introducing these heuristics is to demonstrate the feasibility of systematizing maintenance decisions in a way that balances risks to Ao and ESL. As presented, the heuristic decision trees omit considerations of cost effectiveness or the net value of maintenance. These omissions reflect our understanding that the underlying data needed to support such decisions are not currently available. Such considerations would hopefully be included in subsequent better-developed decision tools.

A Strategy for Prioritizing Maintenance Within SRAs

For the most part, maintenance choices for SRAs seek to balance the overlapping issues of risk and cost. There are potential risks to personnel safety, of near-term loss of availability (put another way, the risk of future opportunity costs), risks that deferred maintenance will increase future maintenance costs, risks that resources available in an availability will not support needed maintenance, risks that deferred maintenance will compromise ESL, and so on. Less recognized are risks that maintenance will have little or no benefit or not be cost effective. There is a near certainty, discussed previously, that issues (such as tank corrosion) ignored in one availability will be significantly more expensive in a subsequent availability. Potential new maintenance concepts introduced here explicitly recognize the issues of risk and cost. Equally important, the concepts would systematize and rationalize necessary maintenance choices and bring better balance to decisions affecting the trade between Ao and ESL.

Absent failure-rate data or data reflecting the duration of maintenance benefits, maintenance costs, or net benefits of maintenance, this study cannot provide detailed maintenance plans to balance Ao and ESL or to conduct maintenance within projected budgets. Under these limitations we offer a heuristic decision tree (shown in Figure 5.1) that could be used to manage Ao and ESL risks. The methodology described here could apply across multiple ship classes.

This decision tree has six decision points labeled A through F. Any systematic treatment of maintenance choices should begin with the Navy's highest priority: personnel safety. This is decision point A. Any maintenance item which, if ignored, would create a significant safety hazard must have the highest priority in maintenance. There is no room for debate here.

If a system does not directly relate to safety, the next decision point (B) is the predictability of failures. Are wear-out failures expected, or will failures be random? Also, for sys-

Figure 5.1
Heuristic Decision Tree for CNO Availabilities

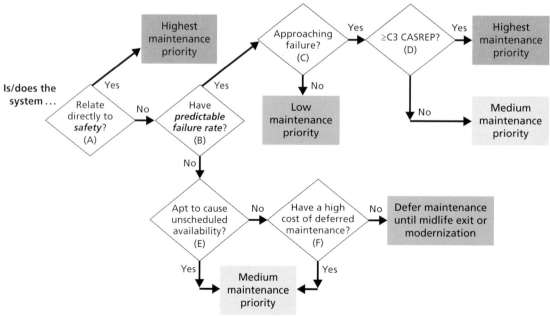

tems with wear-out failure patterns, are those patterns sufficiently well understood, and do *observable* data support a decision? If the failure of a system, such as a diesel engine, can be predicted from wear-out history and indications of impending failure are observed, then the next question (decision point C) is whether the system is approaching failure. If system failure is predictable but, like a diesel engine at mid-life, the system is not approaching failure, maintenance should be a low priority. If a system is approaching failure, then (decision point D), the criticality of the failure becomes a factor. Here criticality is judged on the basis of mission-degrading casualty reports (CASREPs). C3 and C4 CASREPs indicate that a deficiency exists in mission-essential equipment. An equipment deficiency that causes a major degradation, but not loss of a mission-area capability, is designated C3. An equipment deficiency that causes loss of a mission-area capability is designated C4. High priority should be given to maintaining mission-critical systems known to be approaching failure and whose failure would cause a major degradation or loss of mission capability. A decision by SUBMEPP *not* to maintain a mission-critical system known to have a reliability problem is described in Appendix B. Also, impending failures of systems that are not mission critical should have lower priority (i.e., have medium priority) than systems whose failures would lead to major or complete loss of mission capability.

For systems with unpredictable (random) failures, the consequences of failures enter into decision point E. Random failures that would be apt to force an unscheduled availability (especially a dry docking) should be given a medium priority in a CNO availability. An example of this is the stave cables of the SQS-53 sonar, which can fail after ten years of service life and which (currently) can only be replaced in dry dock. In addition to considering the likelihood of failure prior to the next CNO availability, the question of how badly failure would degrade performance (how badly would the loss of a single stave degrade sonar performance?) and the mission implications of a failure. In this example, if anti-submarine warfare capability will be mission critical in the ship's next deployment, a medium level of maintenance should be assigned. Otherwise (an unpredictable failure apt to cause an unscheduled availability is unlikely), the cost deferred maintenance should be considered (in decision point F). Maintenance items with a high cost of deferred maintenance should be given a medium priority. For example, deferring tank maintenance excessively can significantly increase cost when corrective maintenance is ultimately conducted. Decision point F, by explicitly considering the potentially high cost of deferred maintenance, rebalances the maintenance decision process toward lowering risk to ESL.

Heuristics, by definition, use loosely defined rules, and some subjective judgments are required to implement this heuristic. Some decisions points may not have clear "yes or no" answers—"possibly" might be the best answer, with resulting priorities such as medium to low. At this level, we also expect decisions to be informed by considerations not strictly on a given decision branch. A better-defined decision process should be achievable with better data.

The believed reasonableness of the decisions in the heuristic decision tree does not assure the reasonableness of the overall process. To test the reasonableness of the heuristic decision tree, we applied it to five selected SWLINs taken from SRA 3-1 of the DDG-51 TFP. The selected SWLINs are those listed below:

- SWLIN 12311–12321: Tank and Void (T&V) maintenance execution includes inspections (to include open/clean/gas-free), projected structural repairs, and total preservation work. The TFP has identified 15 man-days are necessary for assessments.

- SWLIN 2513X: Condition-based repair and preservation of the propulsion-combustion air intakes and louvers. Mandatory maintenance requirements are for the inspection of intakes annually. SWLIN 2513X tasks can be conducted in CMAVs or CNO availabilities. The TFP has identified 331 man-days are necessary for this maintenance.
- SWLIN 2411X: Mandatory maintenance requirements call for inspection of both main reduction gears (MRGs) prior to each CNO availability. SWLIN 2411X tasks can be conducted in CMAVs or CNO availabilities. The TFP states that 19 man-days are needed for miscellaneous condition-based repairs to the MRG and associated components (main thrust bearing assembly).
- SWLIN 3211X/3241X: Repair of 60 Hz load centers, breakers, switchboards, and cables are condition based and identified through mandatory inspections and mandatory thermal image surveys (every 12 months). Cableway inspections are required every 27 months. SWLIN 3211X/3241X tasks can be conducted in CMAVs or CNO availabilities. Cableway inspection and condition-based repair require 202 man-days.
- SWLIN 46353: The stave cables of the SQS-53 sonar set have an estimated ten-year service life and previously could only be replaced while in dry dock, which required replacement in every DSRA. SHIPALT (S/A) DDG 0051-76231 provides the capability to execute conditioned-based maintenance. Completion of S/A 76231 allows for waterborne replacement of the stave cables when required. This SHIPALT requires 508 man-days.

The heuristic decision tree was followed in assigning priorities for these SWLINs with intermediate steps shown in Table 5.1.

Table 5.1
Intermediate Steps in Prioritizing Sample SWLINs for SRAs

Decision Point SWLIN(s)	(A) Safety?	(B) Predictable Failure?	(C) Approaching Failure?	(D) C3 CASREP?	(E) Unsched Avail?	(F) High Deferral Cost?	Priority
12311–12321: Tanks and voids (15 man-days)	No	No	N/A	N/A	No	Yes	Medium
2513X: Intake plenum repair and preservation (331 man-days)	No	Yes	No	No	No	Yes[a]	Med-Low
2411X: MRG inspection (19 man-days)	No	No	N/A	Yes[b]	Yes	N/A	High
3211X/3241X: Cableway inspection and repair (202 man-days)	Yes	N/A	N/A	N/A	N/A	N/A	Highest
SWLIN 46353: Sonar staves (508 man-days)	No	No	N/A	Possibly	Yes	Possibly	Med-Low

[a] The possibility of damage to gas-turbine engines and/or generators as a result of not maintaining intake plenums informed the assigned priority. Also, projected cost growth as a result of deferring intake plenum maintenance informed the assigned priority.
[b] The possibility of a C3 CASREP as a result of not conducting an MRG inspection informed the assigned priority.

The assessment that cableway inspection and repair is a safety issue was subjective; the assigned priority would have differed otherwise. The priority assignment for SWLIN 2513X (intake plenum repair and preservation) was informed by the prospect of damage to gas-turbine engines or to gas-turbine generator sets. These are events that could cause a C3 CASREP with expensive corrective maintenance. This priority assignment was also informed by the recognition of maintenance cost growth resulting from maintenance deferral. The priority for SWLIN 2411X (MRG inspection) was similarly informed by prospect of MRG or MRG bearing damage (with loss of Ao).

The resulting prioritization, ranked highest to medium-low, is shown in Table 5.2, with a description of potential consequences of deferral.

Table 5.2
Sample SWLIN Prioritization

Priority and Potential Cost SWLIN(s)	Priority	Potential Cost
3211X/3241X: Cableway inspection and repair (202 man-days)	Highest	Greater incidence of grounds—potential safety hazard due to insulation breakdown
2411X: MRG inspection (19 man-days)	High	MRG or MRG bearing repair—loss of Ao
12311–12321: Tanks and voids (15 man-days)	Medium	Additional corrosion; will eventually fail. Cost growth in future.
SWLIN 46353: Sonar staves (508 man-days)	Med-Low	Possible unscheduled dry docking to correct
2513X: Intake plenum repair and preservation (331 man-days)	Med-Low	Could damage gas-turbine engines and generators. Cost growth in future.

Note that this prioritization, with one or more of the lowest-priority SWLINs deferred, would significantly reduce the man-day requirements and cost of SRA 3-1. Also, by giving tanks and voids mid-level priority, a balance between Ao and ESL is seen in the prioritization.

A Strategy for Choosing Between SRAs and CMAVs

As previously noted, CMAVs are the only type of availability accomplished on ships outside of CNO availabilities for non-emergent maintenance. CMAVs can be scheduled or unscheduled; normally last two to six weeks; and are intended for inspections, condition-based upkeep, and minor repairs. CMAVs provide maintenance teams with flexibility in maintenance. Some complex CMAV activities (such as hull, machinery, electrical, electronics, piping work, and ship alterations) are contracted out competitively. Ship's company can conduct other CMAV activities (such as painting or inspections). CMAVs therefore represent a flexible and economical means to accomplish inspections, maintenance, and repairs pierside. With 91 percent of maintenance man-days assigned to maintenance plans going to CNO availabilities, CMAVs account for 9 percent of maintenance man-days. CMAVs therefore present an opportunity to accomplish maintenance that could not be accomplished in SRAs (due to time or fiscal limitations) or to reduce overall maintenance costs.

As with the heuristic for maintenance choices for SRAs, a heuristic for reassigning maintenance from SRAs to CMAVs is presented in the form of a decision tree (shown in

Figure 5.2). Also as before, the methodology described here should be applicable across multiple ship classes.

The heuristic decision tree has four decision points labeled A through D. It begins with a pivotal decision: does the maintenance require in-plant equipment or a dry dock? This is decision point A. If so, the maintenance automatically requires an SRA for completion.

If in-plant equipment or dry docking are not required, the next decision point (B) relates to work required in preparation for maintenance. If there is an excessive requirement to remove interferences or other equipment in preparation for the maintenance, this work should be done in an SRA. Otherwise, the next decision point (C) considers man-days required for the maintenance. If maintenance can be accomplished with a single CMAV with available personnel, consideration should be given to accomplishing it in a CMAV. Otherwise, can the maintenance be broken up into manageable segments (decision point D)? If so, consideration should be given to accomplishing the maintenance in multiple CMAVs.

As with the SRA heuristic, the CMAV heuristic was tested against select SWLINs from SRA 3-1 with intermediate steps shown in Table 5.3.

Figure 5.2
Heuristic Decision Tree Reassigning Maintenance from SRAs to CMAVs

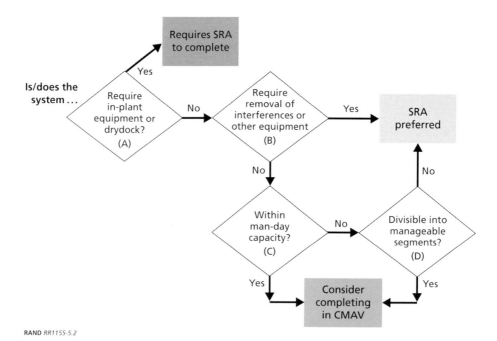

RAND *RR1155-5.2*

Table 5.3
Intermediate Steps in Prioritizing Sample SWLINs for CMAVs

Decision Point SWLIN(s)	(A) In-plant equipment or dry-dock?	(B) Requires removal of interferences or other equipment?	(C) Within man-day capacity?	(D) Divisible into manageable segments?	Maintenance Type
12311–12321: Tanks and voids (15 man-days)	No	No	Yes	N/A	CMAV
2513X: Intake plenum repair and preservation (331 man-days)	No	Yes	No	Yes	Multiple CMAVs, depending on amount of interference required to be removed
2411X: MRG inspection (19 man-days)	No	No	Yes	N/A	CMAV
3211X/3241X: Cableway inspection and repair (202 man-days)	No	No	No	Yes	Multiple CMAVs
SWLIN 46353: Sonar staves (508 man-days)	Yes	N/A	N/A	N/A	SRA

We conclude that only SWLIN 46353 requires an SRA; the other four sample SWLINs could be performed in one or more well-executed CMAVs. We note that all of the SWLINs identified here as suitable for CMAVs are also identified in the DDG-51 TFP as suitable for CMAVs.

This suggests that significant SWLINs that cannot be conducted in SRAs could be reassigned advantageously to CMAVs to accomplish more maintenance or reduce maintenance costs.

Summary for Potential New Maintenance Strategies

SURFMEPP has developed a comprehensive process for achieving Ao and ESL for surface ships through requirements established by technical warrant holders. Historically, DDG-51 maintenance has not paced the stated requirements and cannot be expected to do so in the future. More generally, with observed per-ship annual maintenance costs rising rapidly, the goal of achieving 100 percent of required maintenance will be unachievable. Choices determining which maintenance items will be performed and which will be deferred have been and will continue to be made. These choices have historically favored Ao to the possible detriment of ESL. Cost/benefit data needed to inform surface ship maintenance decisions do not yet appear to exist. The heuristics provided here for selecting maintenance items to defer in SRAs or to move to CMAVs were developed within available data to bring better balance to Ao and ESL and to reduce maintenance costs. These heuristics, which apply across multiple ship classes, can be improved with better data, especially data relating to the cost effectiveness of maintenance items.

Finally, we observe that decisionmakers need to have a better understanding of the term *cost* in making maintenance decisions. Cost can be measured in direct costs of performing maintenance. It can also be measured as the cost to mission capabilities as a result of loss of system availability. Finally, there can be escalating costs (resulting from a so-called fester factor) of deferring some forms of maintenance.

CHAPTER SIX

Conclusions and Recommendations

Conclusions

Maintenance has historically been a fixed fraction of total O&M budgets. Expenditures per ship on surface vessels have varied over the years, not showing a particular trend of steady upward movement. Indeed, in the early 1990s, these expenditures actually declined as the Navy decommissioned whole classes of old ships while building new ones. Since 1998, however, the expenditures per ship have gone up steadily, with an exponential function being the best fit for describing cost per ship through this period. This runs directly into the known legislative environment, which will in fact impact available resources, with an $8 billion annual shortfall between requested budget and legislatively authorized possible if legislation does not change. This is based on a simple comparison between BCA caps and the contents of the Navy Future Years Plan.

Even absent sequestration, continued growth in per-ship maintenance cost is likely unsustainable, at least at the rate seen in the last 15 years. At the rate seen, maintenance would either become a larger component of the O&M budget, come at the expense of new construction or modernization, or require deferral. Maintenance choices, now being made, will become ever-more critical. At a minimum within availabilities, choices would need to be made between activities that favor immediate to favor near-term Ao at the possible expense of longer-term maintenance to protect ESL.

Although we initially attempted to account for the reasons for particular variations among ships, we found that this was a far more complicated problem than existing data would allow us to answer. We did not attempt to account for the differences between ships or explain all the factors that might be impacting a particular trend line. Potential reasons range from training shortfalls to issues with the maintenance infrastructure to deficient manning models. The Navy's SURFMEPP is a relatively new initiative that is collecting ship-level detailed data, and this should prove valuable. In many cases, however, understanding why a particular availability escalated in cost would require on-site interviews with the ship's personnel, maintenance supervisors, and the overall chain of command. This would likely be very worthwhile work, but it rose beyond the sponsor's expectations for the study and is not necessary to answer some of the major issues the study did uncover. Evidence in some ship classes suggests that maintenance deferral places additional pressure on maintenance requirements and thus creates a perhaps insurmountable barrier. Without modification in this path, either some ships will not be maintained at all, or all ships will be maintained at less than the required level. On the current trend, within five years EDSRAs for some DDGs could be deferred by four years with expected negative consequences to cost and readiness. The Navy has historically managed maintenance costs by retiring old, cost-intensive platforms. With pressure to maintain the force size, no new ship classes projected, and with current shipbuilding plans, no such relief for maintenance costs is foreseeable. The surface force the Navy will have in 15 years is substantially the force it has now.

The Navy has attempted to deal with these issues with organizational changes, such as SURFMEPP, and these eventually will likely have a positive influence. But the fundamental decision will still rest at the fleet- and type-commander level and will involve a weighing of operational, materiel, and long- and short-term factors.

A look at the DDG-51 class suggests a mismatch between what the Navy claims is essential for maintenance and what it has historically spent, both in individual years and cumulatively. The Navy publishes TFPs to state the level and kind of maintenance that needs to be performed at regular intervals within successive availabilities, leading to an expected cumulative level of maintenance. We examined in detail the cumulative maintenance levels for DDG-51s and compared them to the levels specified in the TFP. The Navy is not in general funding to the TFP level.

It is not yet possible to assess the effect this mismatch has had on the class—the history is still being written—but there is doubt about either: (1) the validity of the TFP requirement or (2) the Navy's commitment to actually carrying out the maintenance stated in the TFP. In any case, it appears that the Navy will need to consider alternatives to the TFP process as it formulates requirements and resource plans.

Maintenance requirements, as stated in TFPs, do not consider risk in its various forms. They also do not consider the complex issues, including cost, associated with deferring maintenance. They do not consider the net value of maintenance (the duration and degree of improved reliability as a result of maintenance). All of these considerations could contribute in planning more cost-effective ship maintenance.

We also looked at other selected ship classes to evaluate the effect of deferrals, specifically to determine if there is a cumulative effect from deferring maintenance that might not yet be observable with DDGs. The individual histories of ships affected the results. Individual cases suggest, however, that maintenance deferrals exact an extremely high premium on maintenance cost. Equivalent ships of similar age and operating histories whose major difference is basing history—with the attendant effects on maintenance—can show dramatic differences in the overall costs to maintain. Any maintenance construct needs to understand and budget for the high cost of deferral or devise mitigations for cases where deferral is inevitable.

Any mitigation strategy must directly address deferral. Some items may be less critical than others, and not every deferral necessarily carries dire consequences. There is considerable evidence, however, that ships with significant deferral history ultimately become more expensive over the ship's lifetime. Deferral decisions need to be made based on actual materiel condition, actual likelihood that the system will need to be used, and the potential cost of not performing the maintenance. SURFMEPP may ultimately provide some very useful information on materiel history and failure modes to improve ability to predict failures. The decision about whether something is needed and whether the short-term savings outweighs the potential long-term cost, however, is very much a discussion between the fleet and the NAVSEA technical authority.

Data available to the surface community today permits heuristic decisionmaking (with loosely defined rules supported by partial data). This study has presented two such tools in the form of decision trees. The first tool addresses maintenance choices within SRAs. It considers risks and the costs of deferring maintenance and provides a mechanism for evaluating potential work items. The second tool considers whether work-package items might be possibly completed during CMAVs as opposed to longer and more extensive SRAs.

Recommendations

We recommend that the Navy adopt a prioritization scheme that balances cost and risk of deferral against potential near-term needs. The prioritization scheme we have developed balances SRA work items between Ao and ESL and includes a better spread of assignments to CMAVs. For CMAVs to actually achieve efficiencies, it will take dedicated effort by regional maintenance centers, type commanders, contractors, and a ship's force to assure that availability schedules are realistic, maintenance work force has the appropriate size and skill, and complete coordination has taken place prior to the ship beginning the availability. If these are not present, CMAVs might not in fact be more effective. Assuming these conditions can be met, we recommend that the Navy consider adopting these schemes or a variation of them to accomplish the same end.

We also recommend that SURFMEPP follow in the wake of SUBMEPP in better understanding failure modes and patterns, risk, the net value of maintenance, and the costs of deferring maintenance. The SUBMEPP has demonstrated repeatable and beneficial processes for making maintenance choices. As a far more mature program, aspects of SUBMEPP can serve as a model for SURFMEPP.

Maintenance requirements, as stated in TFPs, do not consider risk in its various forms. They also do not consider the complex issues, including cost, associated with deferring maintenance. They do not consider the net value of maintenance (the duration and degree of improved reliability as a result of maintenance). All of these considerations could contribute in planning more cost effective ship maintenance. The Navy should request an independent review of the TFPs by an agency familiar with ship-maintenance practice, but not necessarily tied to the existing Navy practice.

Age and Reliability

The concepts of predictable and random failures were introduced in Chapter Four with reference to OPNAVINST 3000.12A, which uses them to differentiate between Ao calculations for electronic systems and for HM&E systems. This appendix presents an expanded and more nuanced treatment of these concepts and illustrates their use by the Naval Sea Systems Command activity SUBMEPP. We believe that techniques and additional concepts (such as the net value of maintenance) developed and employed successfully by SUBMEPP could be used advantageously in the future by SURFMEPP.

SUBMEPP began generating age and reliability curves using maintenance feedback data in 1998. These curves provided the then-young organization with an objective means to measure the effects of planned maintenance to engineer optimal maintenance plans. The resulting processes were originally applied to *Los Angeles*–class (SSN-688) submarines and remain in use today.[1] As is the case with SURFMEPP, SUBMEPP uses principles from RCM. In doing so, SUBMEPP confirmed the two distinct ways identified in OPNAVINST 3000.12A in which hardware can fail: hardware may wear out (resulting in predictable failures) or may have random (unpredictable) failures.

In more detail, SUBMEPP uses failure-rate curves to categorize casualty-rate patterns with six distinct failure patterns (taken from SUBMEPP and shown in Figure A.1) observed. These graphs depict equipment-failure rates (on the y axis) versus service time (on the x axis). For example, the small Category A reflects systems with high initial failure rates, followed by a period of stability, with a return to high failure rates (perhaps from wear out) late in service life. Category A is thought to be small as a result of improved quality control. Category B reflects systems with a stable low initial failure rate with high failure rates late in service life. Category B might be represented by incandescent light bulbs or hard drives, which rarely fail early in life but have a relatively high failure rate after a known operating period. System failures for these categories are predictable. Category E, the largest category with 56 percent of all systems, has failure rates independent of time. Examples of Category E systems include electronic systems, which tend to fail randomly. Failures for Category E are inherently unpredictable.

The origins of these graphs go back across multiple studies conducted over nearly 50 years. In this period, Category E (constant) failure rates tripled. During the same time period, the occurrence of Category F (infant mortality [that is, early failures]) failure rates has been reduced by 50 percent. One possible explanation (noted by SUBMEPP) for more common Category E failure rates is the increasing prevalence of electronic devices (which can operate for

---

[1] Allen, 2001.

Figure A.1
Observed Categories of Casualty Rates

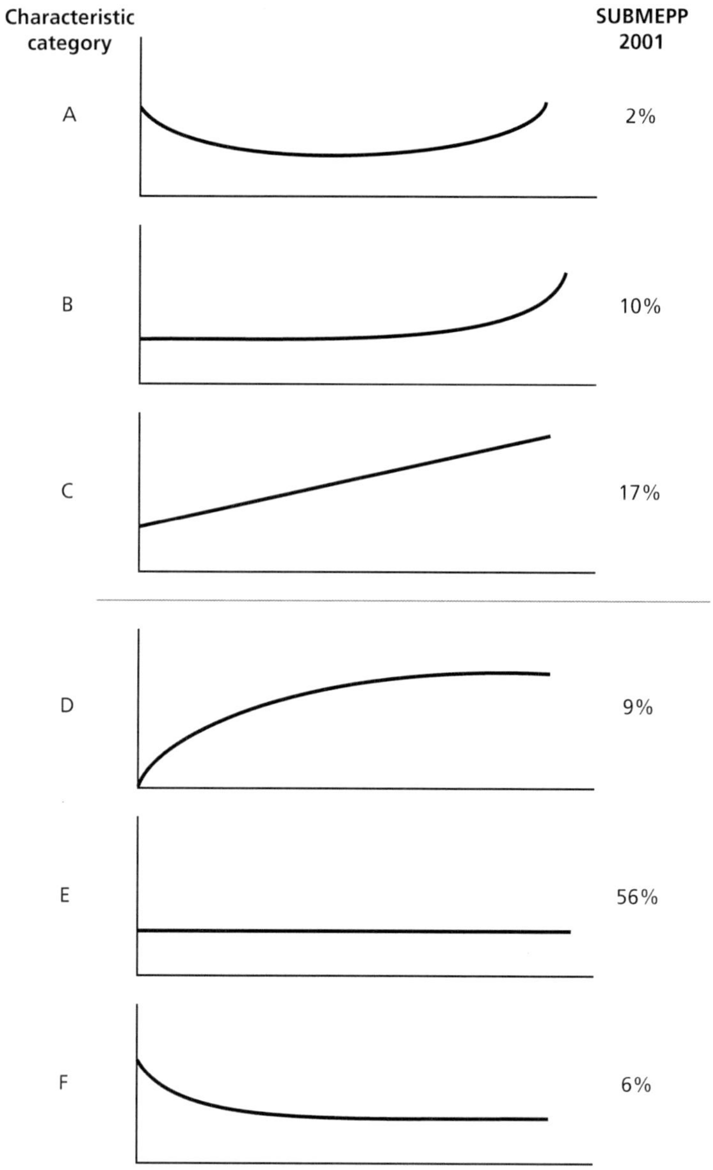

Characteristic category		SUBMEPP 2001
A		2%
B		10%
C		17%
D		9%
E		56%
F		6%

RAND *RR1155-A.1*

years then fail without warning). The decrease in infant mortality rates might reflect improving quality control in manufacturing processes and more extensive acceptance testing.

Some of SUBMEPP's findings are echoed in more recent and independent studies of casualty rates. In particular, the marine-diesel manufacturer MTU has conducted a nuanced study of failure rates using the metric of time between major overhauls (TBO).[2] MTU defines TBO as the time span in which operation without major failure is ensured, precluding wear-

---

[2] MTU Friedrichshafen GmBH, "Technical Project Guide, Marine Application, Part 1-General," revision 1.0, Friedrichshafen, Germany: MTU Friedrichshafen GmBH, June 2003.

related damage (with a 1-percent threshold) requiring a major engine overhaul or replacement (referred to by MTU as Maintenance Echelon W6). As illustrated in Figure A.2, MTU has observed an initial period with an elevated period of early failures followed by a relatively long period of random failures, ending with a period of wear-out failures. SUBMEPP analysts saw, for example, a similar pattern in salvage air valves for the SSN-688 class. A takeaway here is that even systems having terminal wear-out failure periods should be recognized as having long periods in which random failures are the rule.

Figure A.2
MTU Definition of TBO

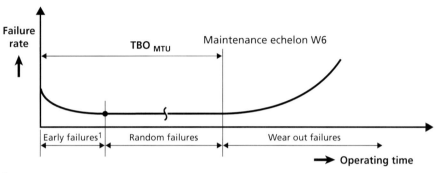

¹Probable start-up failures

RAND *RR1155-A.2*

The net value of maintenance is an important concept in the SUBMEPP process. Figure A.3, adapted from SUBMEPP, illustrates its finding that steady improvements in reliability may be difficult to sustain through additional maintenance. At some point, the majority of items that are going to fail have failed, so maintenance provides little additional reliability. Dramatic failure-rate decreases may be followed shortly by swift returns to the status quo. The shaded area in this figure represents a net improvement in reliability for a system with a planned overhaul interval of five years. The benefits of maintenance here are transitory, and the maintenance itself is not cost effective.

Figure A.3
Possible Reliability Effect with Five-Year Overhaul Cycle

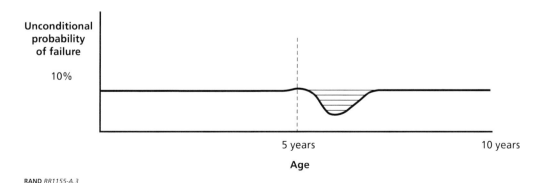

RAND *RR1155-A.3*

Figure A.4
Various Reliability Effects with Five-Year Overhaul Cycle

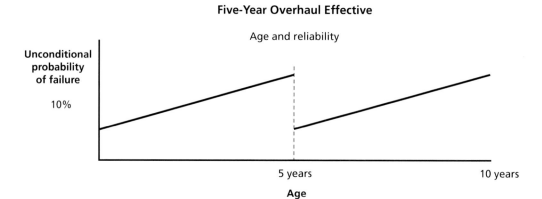

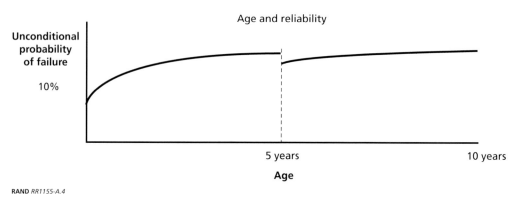

RAND *RR1155-A.4*

The net benefits in reliability for two systems are illustrated in Figure A.4 and also taken from SUBMEPP. Both systems have planned five-year overhaul cycles. The system illustrated in the top portion of Figure A.4 benefits significantly from overhauls, whereas the system in the lower-portion component experiences increased failure rates as it ages, but corrective maintenance does not improve the situation. An air-dehydrator system examined by SUBMEPP illustrates the latter case, in which corrective maintenance provided little value. The air dehydrator experienced increased failure rates as it aged, and corrective maintenance did not improve the situation. As a result of backup capabilities, the failure modes in question did not have significant consequences, however, consideration was given to restoring the unit from a cost-containment perspective. The system engineer went on to compare reliability following past overhauls to the reliability exhibited by a new component and found that little improvement would be gained from overhauls.

Most Significant SWLINs for DDG-51

The LMRS for the DDG-51 currently has a total of 2,368 SWLINs. Forty-four SWLINs or groups of SWLINs are identified in the DDG-51 TFP as having significant life-cycle effect, significant cost drivers, or both. We summarize them here for reference purposes.

SWLIN Series 100, Hull Structure

Maintenance in this series is associated with assessment, repair, and preservation of hull structure. A total of 49,743 (Flt I/II) and 57,995 (Flt IIA) man-days are distributed across their respective ESLs. Within SWLIN Series 100, the items determined to contribute most significantly to life cycle impact and/or significant cost drivers are:

- SWLIN 11011: Underwater hull inspection and repair is to be accomplished in every dry-docking cycle. SWLIN 1101X tasks can be conducted in CMAVs or CNO availabilities.
- SWLIN 11312: Replacement of six of the 18 bilge drain wells is to be synchronized with scheduled tank preservation during every dry docking. The remaining 12 bilge drain wells will be inspected, repaired, tested, and preserved.
- SWLINs 12311–12321: Tank and Void (T&V) maintenance execution includes inspections (to include open/clean/gas-free), projected structural repairs, and total preservation work.
- SWLIN 15011: Structural deterioration of the 02 through 06 level superstructure deck plating requires authorized installation of thicker deck plating. Notional man-day requirements are allocated to each CNO depot-maintenance period.
- SWLIN 16511: Radiographic inspection and minor repairs are to be accomplished to the sonar dome structure and rubber window in every dry docking.
- SWLIN 16711: Maintenance, powder coating, and replacements of exterior watertight doors, hatches, and scuttles are conducted to maintain watertight integrity for ship's survivability and damage control. SWLIN 16711 tasks can be conducted in CMAVs or CNO availabilities.

Each of these SWLINs Series 100 is described as a high-cost driver and Life Cycle Critical (LCC), or as necessary to address and execute to achieve ESL. Every man-day requirement for all of these SWLINs Series 100 is stated to be based on historical trends and maintenance community inputs. No basis for prioritizing these SWLINs Series 100 is evident.

SWLIN Series 200, Propulsion Plant

Work in this series is associated with assessment, repair, and preservation of the propulsion plant. A total of 56,200 (Flt I/II) and 61,853 (Flt IIA) man-days are distributed across their respective ESLs. Within SWLIN Series 200, the items determined to contribute most significantly to life-cycle impact and/or significant cost drivers are:

- SWLIN 2341X: Mandatory maintenance requirements call for inspection of the module resilient mounts for the LM2500 propulsion gas turbines, with a 20-year mandatory replacement interval.
- SWLIN 2411X: Mandatory maintenance requirements call for inspection of both MRGs prior to each CNO availability. SWLIN 2411X tasks can be conducted in CMAVs or CNO availabilities.
- SWLIN 243XX and 244XX: Mandatory maintenance requires the removal, inspection, and repair of propulsion shafts and struts every ten years. Portions of SWLIN 243XX and 244XX tasks can be conducted in CMAVs or CNO availabilities.
- SWLIN 245XX: Mandatory maintenance requirements for the inspection and repair of the Controllable Pitch Propeller every dry docking.
- SWLIN 2513X: Condition-based repair and preservation of the propulsion-combustion air intakes and louvers. Mandatory maintenance requirements are for the inspection of intakes annually. SWLIN 2513X tasks can be conducted in CMAVs or CNO availabilities.
- SWLIN 2591X: Condition-based repair and preservation of the propulsion exhaust air uptakes. SWLIN 2591X tasks can be conducted in CMAVs or CNO availabilities.

Each of these SWLINs Series 200 is described as LCC. All SWLINs (other than SWLIN 2411X, which calls for 19 man-days of labor) are described as high-cost drivers. No basis for prioritizing these SWLINs Series 200 is evident.

SWLIN Series 300, Electric Plant and Distribution Equipment

Work in this series is associated with assessment and repair of the electric plant. A total of 7,305 (Flt I/II) and 8,132 (Flt IIA) man-days are distributed across their respective ESLs. Within SWLIN Series 300, the items determined to contribute most significantly to life-cycle impact and/or significant cost drivers are:

- SWLIN 3113X: Condition-based repairs to gas turbine generator sets are identified through mandatory pre-deployment and nonperiodic assessments. Mandatory (14-year) resilient mount replacement is expected to occur twice during ESL. SWLIN 3113X tasks can be conducted in CMAVs or CNO availabilities.
- SWLIN 3211X/3241X: Repair of 60 Hz load centers, breakers, switchboards, and cables are condition based and identified through mandatory inspections and mandatory thermal image surveys (every 12 months). Cableway inspections are required every 27 months. SWLIN 3211X/3241X tasks can be conducted in CMAVs or CNO availabilities.

- SWLIN 3431X: Generator support-system repairs are condition based and identified by mandatory pre-deployment inspections and nonperiodic assessments. Lube oil cooler cleaning/repair and gas-turbine generator exhaust system repairs are the cost drivers in this standard SWLIN. SWLIN 3431X tasks can be conducted in CMAVs or CNO availabilities.

Each of these SWLINs Series 300 is described as LCC and mandatory, and high-cost drivers. No basis for prioritizing these SWLINs Series 300 is evident.

SWLIN Series 400, Command and Surveillance

The work in this series is associated with assessment, repair, and preservation of command and surveillance systems. A total of 47,753 (Flt I/II) and 46,529 (Flt IIA) man-days are distributed across their respective ESLs. The items determined to contribute most significant life cycle impact and/or significant cost drivers are:

- SWLIN 45511: Repairs, overhaul, and replacement of the identification friend or foe (IFF) antenna group are condition based and identified through inspection. SWLIN 45511 tasks can be conducted in CMAVs or CNO availabilities.
- SWLIN 45601: Repairs to the SPY-1 multimode radar are condition based and identified through mandatory inspections (12- and 24-month intervals). Array resurfacing is condition based and is expected to occur every DSRA.
- SWLIN 46353: The stave cables of the SQS-53 sonar set have an estimated ten-year service life and previously could only be replaced while in dry dock, which required replacement in every DSRA. SHIPALT (S/A) DDG 0051-76231 provides the capability to execute conditioned-based maintenance. Completion of S/A 76231 allows for waterborne replacement of the stave cables when required. This will eliminate the requirement to completely replace cables and transducers every DSRA. DDG 51-78 are planned to have this alteration executed during the EDSRA/Depot Modernization Period unless directed to execute during a DSRA. Flt IIA ships are scheduled to receive S/A 76231 during their next DSRA. Notional man-day requirements to support condition-based repairs are allocated to the applicable CNO depot-maintenance periods.
- SWLIN 47111/47211: Significant costs are associated with SLQ-32 electronic-countermeasures system actuator replacement and cooling-system heat-exchanger repair. Repairs are condition based and identified through mandatory inspections. The allocation of man-days to support repairs is not stated.
- SWLIN 48134: MK34 fire-control system and gun repairs are condition based and identified through inspections. Significant repair costs are attributed to MK46 optical sight system repairs and refurbishment. Notional man-day requirements to support condition-based repairs are allocated to the applicable CNO depot-maintenance periods.
- SWLIN 48299: MK99 fire control system repairs are condition-based and identified through various mandatory inspections (12–72 months). Significant costs are attributed to MK82 director life-cycle repairs and one mandatory replacement during mid-life period. Notional man-day requirements are allocated to the applicable CNO depot-maintenance periods.

Each of these SWLINs Series 400 is described as LCC, mandatory, and high-cost drivers. No basis for prioritizing these SWLINs Series 300 is evident.

SWLIN Series 500, Auxiliary Systems

The work in this series is associated with assessment, repair, and preservation of auxiliary systems and components. A total of 57,396 (Flt I/II) and 76,706 (Flt IIA) man-days are distributed across their respective ESLs. The items determined to contribute most significant life-cycle impact and/or significant cost drivers are:

- SWLIN 50810: Machinery and non-machinery insulation and lagging repairs are condition based and identified through inspection. Notional man-day requirements to support condition-based repairs are allocated to each depot-maintenance period (CMAV and CNO).
- SWLIN 51211 and 51311: Ventilation-ducting repairs commence at approximately ten years into service life and are expected to continue throughout service life. Overhaul of three ventilation-system fans is expected within a 32-month periodicity (FRP) and is planned as part of a condition-based repair strategy. In conjunction with fan and ducting repairs, ventilation system cleaning is planned at ten-year intervals. Notional man-day requirements to support condition-based repairs are allocated to each depot-maintenance period (CMAV and CNO).
- SWLIN 51421: A/C plant compressors are identified as having a high failure rate. The compressor shaft seals have an expected service life of 18 months. A/C plants using HFC-134a Freon® require compressor overhaul approximately every eight years due to increased system pressure. Additionally, condenser cleaning is required every 32 months to maintain system parameters. Notional man-day requirements to support repairs are allocated to each depot period (CMAV and CNO).
- SWLIN 52011: Each DSRA, 80 sea valves are planned for repair or replacement. Unscheduled sea-valve repair using a cofferdam during CMAV periods is also anticipated.
- SWLIN 52111: Deterioration of fire-main piping begins to occur approximately ten years into service life. Repairs are condition based and identified through inspection. Notional man-day requirements to support condition-based repairs are allocated to each depot-maintenance period (CMAV and CNO).
- SWLIN 55151: High-pressure air compressor (HPAC) (Flt I/II only) top-end overhaul is required every CNO availability, and full overhaul of both HPACs is required every third CNO availability. All notional man-days to support condition-based repairs are allocated to CNO availabilities.
- SWLIN 56111: The steering system is reliable with a less than 5-percent annual component failure rate. Due to redundancy built into the design, component failure does not restrict operation of the steering gear. The repair strategy is to overhaul two of the four hydraulic steering pumps every ten years. Also, the steering system has 42 critical flex hoses that have mandatory replacement periodicity of 240 months. Finally, there are 21 noncritical flex hoses that have a 20-year service life. Replacement starts in the second docking cycle and is sequenced to be executed during CMAV periods through end of service life.

- SWLIN 56211: Rudderstock sleeves should be replaced with each scheduled dry docking. Man-day requirements to support repairs are allocated to the appropriate CNO depot-maintenance periods.
- SWLIN 58311: Corrosion severely reduces service life of slewing arm davits, and some of their components are obsolete. Corrosion maintenance is required with man-day requirements to support repairs allocated to the appropriate depot-maintenance periods (CMAV and CNO).
- SWLIN 58821: For Flt IIA only, replace single and tandem helicopter hangar–door rollers every 18 months; replace sealing-flap return springs and water and fire seals every 24 months; remove, clean, and refurbish drive chain and lubricate drive-shaft bearings every 60 months; repair the lift arm assemblies every 72 months; and replace wire rope every 96 months. Notional man-day requirements to support repairs are allocated to the appropriate depot-maintenance period (CMAV and CNO).
- SWLIN 58821: Inspection of the entire recovery assist, securing, and traversing (RAST) system is required every 24 months. Analysis of historical data indicates repairs are required every 27 months in support of aviation certification. Notional man-day requirements to support repairs are allocated to the appropriate depot-maintenance period (CMAV and CNO).
- SWLIN 59311: Repairs to vacuum collection, holding, and transfer (VCHT) piping typically commence approximately ten years into service life. Repairs are condition based and identified through inspection. Also, VCHT piping requires cleaning every CNO availability to maintain the system. Notional man-day requirements to support repairs are allocated to the appropriate depot-maintenance period (CMAV and CNO).

SWLIN Series 600, Outfit and Furnishings

The work in this series is associated with assessment, repair, and preservation of ships structure. A total of 45,740 (Flt I/II) and 48,976 (Flt IIA) man-days are distributed across their respective ESL. The items contributing significant life cycle impact and/or significant cost drivers are:

- SWLIN 63111: Interior preservation. Notional man-day requirements to support repairs are allocated to the appropriate depot-maintenance period (CMAV and CNO).
- SWLIN 63120/63121: Exterior preservation (including masts). Notional man-day requirements to support repairs are allocated to the appropriate depot-maintenance period (CMAV and CNO).
- SWLIN 63131: Underwater body-hull preservation is required at every dry docking.
- SWLIN 63141: Freeboard preservation is required at every dry docking.
- SWLIN 63311: The impressed current cathodic protection (ICCP) system and sacrificial anodes are high-cost drivers and are LCC. Notional man-day requirements to support these routine work requirements are allocated every drydocking.
- SWLIN 63411: Exterior nonskid deck covering maintenance is assigned to the appropriate depot-maintenance period (CMAV and CNO) to address backlog and to maintain exterior coating systems.

- SWLIN 63421: Flight and hangar-bay deck nonskid coating system maintenance is allocated to the appropriate depot-maintenance period (CMAV and CNO) as required to meet Aviation Certificate (AVCERT) requirements.

References

Allen, Timothy M., *U.S. Navy Analysis of Submarine Maintenance Data and the Development of Age and Reliability Profiles*, Portsmouth, N.H.: Department of the Navy, 2001.

Cavas, Christopher P., "Two Ships Deemed 'Unfit' for Combat," *NavyTimes.com*, April 20, 2008.

Creevy, Rear Admiral Larry, and Rear Admiral William Galinis, interview with authors, Washington, D.C., August 2014.

Department of the Navy, "Fiscal Year (FY) 2014 Budget Estimates, Justification of Estimates, Operation and Maintenance, Navy," April 2013. As of October 14, 2015:
http://www.secnav.navy.mil/fmc/fmb/Documents/14pres/OMN_Vol1_BOOK.pdf

Department of the Navy, Commander, U.S. Fleet Forces Command, "Command Investigation of Diesel Engine and Related Maintenance and Quality Assurance Issues Abroad USS *San Antonio* (LPD 17)," 5830 / Ser N00/131, May 20, 2010.

Elmendorf, Douglas W., letter to John A. Boehner and Harry Reid re Budget Control Act of 2011, Washington, D.C., August 1, 2011.

———, "Estimated Impact of Automatic Budget Enforcement Procedures Specified in the Budget Control Act," Washington, D.C., September 12, 2011.

Larter, David, "Three-Star: Surface Fleet Readiness, Training Are on Track," *NavyTimes.com*, January 13, 2015.

Malone, Michael, "Surface Force Maintenance Engineering Planning (SURFMEPP)," project briefing given to the Surface Navy Association (SNA), January 17, 2013.

MTU Friedrichshafen GmbH, "Technical Project Guide, Marine Application, Part 1-General," revision 1.0, Friedrichshafen, Germany: MTU Friedrichshafen GmbH, June 2003.

Naval Sea Systems Command, "Technical Foundation Paper for LPD 17 Class Revised Type Commander Notionals," Washington, D.C.: U.S. Navy, 2011, p. 5.

Nemfakos, Charles, former assistant secretary of the Navy, Financial Management and Comptroller, interview with authors, Washington, D.C., June 15, 2014.

New Wars, "The Balisle Report and the Navy's Future Part 1," July 19, 2010. As of September 22, 2015:
https://newwars.wordpress.com/2010/07/19/the-balisle-report-and-the-navys-future-pt-1

Office of the Chief of Naval Operations, *Maintenance Policy For United States Navy Ships,* OPNAV Instruction 4700.7L, Washington, D.C.: Department of the Navy, May 25, 2010.

———, "Operational Availability Handbook: A Practical Guide for Military Systems, Sub-Systems and Equipment," enclosure in *Operational Availability of Equipments and Weapons Systems,* OPNAVINST 3000.12a, N40, Washington, D.C.: Department of the Navy, September 2, 2003.

———, *Optimized Fleet Response Plan*, OPNAVINST 3000.15, USFF/CNO N3/N5, Washington, D.C.: Department of the Navy, November 10, 2014. As of September 24, 2015:
http://doni.daps.dla.mil/Directives/03000%20Naval%20Operations%20and%20Readiness/03-00%20 General%20Operations%20and%20Readiness%20Support/3000.15A.pdf